*f*P

THE PERFECT ENGINE™

How to Win in the New Demand Economy
by Building to Order with Fewer Resources

ANAND SHARMA
PATRICIA E. MOODY

THE FREE PRESS
New York   London   Toronto   Sydney

THE FREE PRESS
A Division of Simon & Schuster, Inc.
1230 Avenue of the Americas
New York, NY 10020

Copyright © 2001 by Anand Sharma and Patricia E. Moody
All rights reserved,
including the right of reproduction
in whole or in part in any form.
The Perfect Engine is a trademark of TBM Consulting, Inc. All rights reserved.
LeanSigma is a service mark of Maytag Corporation, used under exclusive license to
TBM Consulting, Inc. All rights reserved.
Any other trademarks or service marks referenced in this book are the property of their
respective owners.

THE FREE PRESS and colophon are trademarks
of Simon & Schuster, Inc.
For information about special discounts for bulk purchases, please contact
Simon & Schuster Special Sales: 1-800-456-6798 or business@simonandschuster.com

Designed by Deirdre C. Amthor

Manufactured in the United States of America

10 9 8

Library of Congress Cataloging-in-Publication Data

Sharma, Anand, 1945–
The perfect engine : how to win in the new demand economy by building
to order with fewer resources / Anand Sharma, Patricia E. Moody.
p. cm.
Includes bibliographical references and index.
1. Manufacturing industries—Management. 2. Manufacturing industries—
Technological innovations. 3. Production management. 4. Manufacturing processes.
5. Industrial management. I. Moody, Patricia E. II. Title.

HD9720.5 .S27 2001
658.5—dc21   2001040286
ISBN 978-1-4516-4085-4

The authors gratefully acknowledge permission from the following source to reprint material in their control:

Maytag Corporation for the photo of the Maytag Repairman. Copyright © 2001 by Maytag Corporation.

For Anu, Allie, and Drew
Anand Sharma

For all The Mill Girls
Patricia E. Moody

Contents

ACKNOWLEDGMENTS		ix
FOREWORD BY LLOYD WARD, FORMER CEO OF MAYTAG CORPORATION		xi
PREFACE: TODAY'S MANUFACTURING CHALLENGE		xv
Chapter 1	A Better Way	1
Chapter 2	WHACK! What Doesn't Work and Why	34
Chapter 3	LeanSigma Leadership	56
Chapter 4	Preparation for Transformation and Innovation	79
Chapter 5	Lean Production System	108
Chapter 6	Lean Ergonomics and Safety	147
Chapter 7	Design for LeanSigma	173
Chapter 8	Maintaining the Gains in a Culture of Change	200
Chapter 9	The Value Chain	237
Chapter 10	The Future	256
LIST OF FIGURES		265
BIBLIOGRAPHY		271
INDEX		273
ABOUT THE AUTHORS		281

Acknowledgments

Every book is a timeless accumulation of facts and ideas, and inspiration, and hope. We are grateful to the hundreds of individuals and companies who provided so many facts, and so much hope.

The Maytag Corporation, especially Lloyd Ward, former CEO; Carole Uhrich, former president of Maytag Home Solutions; Art Learmonth, VP Mfg. and Engineering of Maytag Home Solutions; Thomas Schwartz, VP Communications; Tom Briatico, VP and general manager; and Ramin Zarrabi.
Pella Corporation: Gary Christensen, CEO; Mel Haught, executive vice president; and Brian Giddings.
Mercedes-Benz do Brasil: Karsten Weingarten.
Wiremold: Art Byrne, president and CEO; and Orest Fuime.
Lantech: Pat Lancaster, founder and chairman; Ron Hicks, VP; and Jean Cunningham, CFO.
Padilla, Speer & Beardsley: Mike Greece, for creative input.
Triangle Orthopedic: Dr. Bill Mallon, for ergonomics information.
Shingijutsu Co., Ltd., Nagoya, Japan: the late Mr. Yoshiki Iwata, chairman and CEO, for Lean Production System teachings and a decade of inspiration.
TBM Consulting Group: Sam Swoyer, Bob Dean, Mark Oakeson, Mike Herr, Dan Sullivan, Stephen Smith, and all the consultants

Acknowledgments

who worked relentlessly transforming and creating hundreds of examples covered in this book. Kathy Million, Emily Adams, Mary Mann, Rich Houghtaling, and Jennifer Rainey, who provided the administrative support, and especially Lisa Truitt White, who helped with all the artwork.

Cindy Slater of Maytag Corporation, and dozens of manufacturing associates provided valuable insight and inspiration.

Foreword
By Lloyd Ward
Former CEO, Maytag Corporation

Imagine what it must have been like one hundred years ago for business leaders and entrepreneurs of the day. Standing at the dawn of a new century, looking at the profound changes that were unfolding in society and in commerce . . . absorbing the brash challenges to conventional wisdom . . . calculating the untold opportunities change would create . . . *if* those individuals were bold enough to grasp the opportunities . . . *if* those individuals were bold enough to invent the future in the new century.

And what would those same individuals say to us today . . . in our place . . . at this time . . . as we look ahead to the profound changes that are unfolding early in this century . . . to the unlimited opportunities that change will offer . . . *if* we are bold enough to challenge conventional wisdom . . . *if* we are bold enough to grasp opportunity . . . *if* we are bold enough to invent the future.

Transformation is key. This book is about being bold enough to invent the future. To transform the present into the future. It's about more than the change required to meet the future in business; it's about transforming to invent the future in business. Change without transformation is like breathing air without oxygen: the lungs are working but it's not very nourishing. For those of us who lead "old economy" companies, our challenge is to invent the future—grounded in the capabilities of value creation in the old economy—with sights set on the possibilities of value creation in the new economy.

Foreword

Maytag's journey of transformation is taking us further into familiar territory than we had ever dreamed . . . and deeper into unfamiliar territory than we had imagined possible—or necessary. In that transformation, we are determined to invent a future that is compelling for investors, employees, and consumers.

How?

Maytag's alchemy of old economy capabilities and new economy possibilities is straightforward: Deliver the heritage business in a new century. Expand globally. Extend digitally. Transform completely. Building a strategy of Intelligent Innovation to deliver our heritage business in a new century, we have chosen a strategy of Intelligent Innovation behind premium brands—Maytag, Hoover, and Jenn-Air. Intelligent Innovation is how we describe the intersection of what is technologically feasible from the engineer's point of view and what is technologically desirable from the consumer's point of view. Finding the intersection can be magical. It is the point where we can touch consumers with products and services that ease the work of life and enhance the joy of living.

Delivering Intelligent Innovation behind heritage brands in a new century requires quality and productivity measured against world standards, best captured for us by the LeanSigma Transformation. LeanSigma creates operating excellence, by combining the principles of lean manufacturing and the discipline of Six Sigma quality.

Delivering Intelligent Innovation behind heritage brands in a new century requires putting innovation in front of every consumer in every market on every purchase occasion in which we can create value anywhere in the world.

Delivering Intelligent Innovation behind heritage brands in a new century requires creating underlying value with old economy assets at the same time we explore breakthrough value creation with Web-enabled business models and new economy possibilities. And it requires extending the heritage business into new value creation models. Lifestyle kitchens where a consumer can prepare an entire meal without moving a step; lifestyle vending where at home, at work, and at leisure events specialized "vendors" deliver products and services that consumers want and need at the time; New Age appliances in vehicles such as the Ford Windstar minivan so that consumers can travel with a mobile kitchen and laundry room.

Foreword

We recognize that consumers will continue to be catalysts for change in business; certainly consumers have been catalysts in our focus on Intelligent Innovation. When we touch consumers with Intelligent Innovation, we change the game. The more often we find that opportunity, the more consumers come to expect from our brands and our innovation. The more consumers expect, the more inspired we will need to be with innovation. Sooner or later, consumers will expect customized appliances, and they will expect us to build them to order, with the features they want, and deliver the product when the consumer wants it delivered. We expect to be a leader in this consumer revolution. It's the logical extension of Intelligent Innovation. It's also why we're pursuing LeanSigma so aggressively.

Manufacturing transformation in the new economy, "Five days or less!" We have a vision to take an order electronically and build the product for delivery in five days. To transform our manufacturing process to deliver in five days is more than working harder, faster. That would be like trying to take a 1978 Chevrolet and rework the engine to the point where it will get thirty-five miles to the gallon. There is only so much modification that can be done. At some point you have to rebuild the entire car from the ground up—begin at a different starting point so that you finish at a different end point.

That's what is so impressive about the Lean approach to the supply chain. The transformation begins with hard-edged productivity focus in the manufacturing facility and extends downstream and upstream in the supply chain. Everything that Maytag is doing—its transformation into an Innovation Machine—is part of the story of *The Perfect Engine*. We at Maytag know from experience that LeanSigma, Kaizen, Design for LeanSigma (formerly 2P and 3P), Lean Ergonomics, and all the other wonderful methods uncovered in *The Perfect Engine* are the tools that will take us there. Strategically, we, like so many other companies, want to move fast to improve all our processes, and we know that the lean tools accelerate the process.

A number of Maytag employees have asked whether we really mean that we want to be able to take an order electronically and build product in five days. Absolutely. The vision is to build to order. But it's also more than that. To build to order we will need to improve our manufacturing

Foreword

and supply chain processes in ways we have never dreamed possible. Along the way, every product and every process in every business we operate will benefit. This will represent a pivotal point in our transformation to smartly leverage "old economy" capabilities and "new economy" possibilities. Transformation is the issue, not destroying the old business. Those who invent the future most creatively will be those businesses that retain the strength of their old economy value creation yet add dimensions through which they will reach new economy possibilities.

Leadership at the turn of the previous century faced the future with a boldness of vision . . . a determination to grasp opportunity . . . an unquenchable desire to lift themselves . . . commerce . . . society to full potential. That also is the challenge and opportunity all of us face in this century. This book offers a worldview that celebrates both.

Preface: Today's Manufacturing Challenge

- Today's consumer expects high-quality products and services, delivered on demand, customized to individual taste at a reasonable price.
- Today's employees expect to work in a challenging yet gratifying environment, in which their talents and contributions are recognized and rewarded.
- Today's supply chain partners expect mutually beneficial long-term relationships.
- Today's shareholders expect profitable growth and competitive market valuations.

Is your organization ready to respond?

The challenge in this economy is for manufacturers to ready themselves to respond. They must learn to produce on actual demand, within the acceptable lead time, and to respond quickly to delight their customers. They must use the Internet to provide line-of-sight to customer desires throughout their value chain, and they must extend it to their supply chain instantly.

They must learn how to make many things in different places—

Preface

Distributive Manufacturing. They must design and deliver product that is mass customized and personalized, at the same cost that prevailed under mass production economy-of-scale methods.

This book is written for the producers who want to satisfy all of their stakeholders; grow their companies three to five times the industry growth rate, at twice the rate of profitability, and understand Distributive Manufacturing. Fortunately, there are a few glowing examples of manufacturers who have become lean and responsive, who have extended their power and speed through careful application of the LeanSigma Transformation Process. They know how to cut lead times and perfect processes that deliver perfect product, on time, at globally competitive prices. Not only that, but the cultural model of these "process" leaders portends highly successful lessons for the highly demanding digital economy ahead. We call the system that they have built The Perfect Engine. It works.

THE PERFECT
ENGINE

CHAPTER 1
A Better Way

The Grind

Stepping into the huge kitchen cabinet assembly plant, you are assaulted by the sights, sounds, smells, and by-products of a very busy operation. There is a thick haze of sawdust in the air and on the floor. Mile-high racks of parts storage hold an accumulation of dusty laminated doors and trim pieces. A fleet of fork trucks race down the aisles, moving empty bins and depositing fresh crates of piece parts in open areas that become acres of in-process inventory storage.

Your eyes, stung by paint and glue fumes, burn, and you start to sniffle and sneeze as the vapors and dust settle in on your clothing. Tiny bits of particulate matter float by.

At shift change, operators blow the sawdust off their equipment with air hoses; the material makes fine grit underfoot until hours later, when a sweeper comes by to stir up new piles of sawdust and pieces of laminate. He works his way through the plant, pushing and piling mountains of accumulated trimmings—evidence of yesterday's, and last week's, and last month's endless attempt to make schedule. Please customers. Fill trucks. Get paid.

Out at the shipping dock, trucks appear hourly to unload heavy sheets of plywood and laminate. Suppliers hustle boxes of hardware and drawer fixtures while shipping clerks, overwhelmed with the press of paperwork

THE PERFECT ENGINE

and fork trucks and upstream demands for more raw material, move from one disconnected operation to another.

On the floor, fork trucks rush pallets of raw material to cutting machines; the big saws' high whine makes it impossible to understand operators' shouted explanations of their process. Everything about this plant is busier, noisier, dirtier, and heavier than what one would expect from a twenty-first-century North American manufacturing giant.

This particular building houses final assembly for a high-volume producer of premium wood cabinetry. It's a complex operation and, with a booming construction economy fueling strong demand, every day is an opportunity.

The best way to understand the scope and rhythm of any facility is to follow one complete product from receipt of raw material down to various processing steps, into final assembly, packaging, and the shipping dock. This plant, however, presents a special challenge because its multiple subassembly and processing departments feed huge variety to the final assembly lines. It is possible to walk through key subassembly areas that feed the final assembly line, and each one of them is an eye-opener.

Down on the white door line a team is tackling one gigantic lamination machine that seems to stall out once per shift. The work stoppage ripples outward and causes immediate downstream disruption as four expeditors from final assembly converge on a lone table saw operator. Larry is a six-year veteran of endless rush orders, expedites, and firefighting. The expeditors are impatient and they wave scraps of paper bearing endless parts shortage lists in his face—"line's down," "gotta have it," "can't find it," "big customer," and "won't wait" punctuate their demands.

Confronted with four orders for hot shortages, Larry silently moves to his small work cell and begins, one by one, to cut parts. There's a quiet determination about him that belies the hopelessness of his task. Every day, Larry's work routine becomes a long series of interrupted and equally frantic calls for help from downstream assembly workers who cannot keep their lines running, who must pull incomplete cabinets off to the side while they wait for missing pieces.

In fact, what was designed to be a smooth line-of-sight assembly has

A Better Way

been transformed by a nightmare process filled with missing doors and damaged trim pieces into a line interrupted, a broken series of incomplete customer orders. Everything waits; nothing flows. And yet, final assembly is where all the sins, all the missed deliveries and quality issues and design problems make their very visible appearance. While operators can still be expected to work the occasional miracle, they simply cannot run lines with no parts. Henry Ford knew this, countless appliance and electronics and computer factories proved this, and certainly the competitors know this.

It's every operations manager's nightmare, every customer's frustration, and Larry's problem. But this hurry-up-and-wait way of running manufacturing is not atypical—thousands of factories across the world struggle from day to day with the same uneven pace, the same horrific ergonomics and the same frustrated customers.

Throughout the plant there are other signs of a bad operation—an imbalance of huge computer-controlled machines played off against highly used, small, manually operated equipment. Long lines are broken by accumulations of mismatched parts, operators working to keep up a desperate pace, and workers who wander from one operation to another. At the end of the day they return home not knowing exactly what they have produced, or how they may have accomplished some vital piece of the company's mission.

For years customers have ordered semicustom products—oak, maple, or birch cabinets of any size or height configuration—for promised delivery within six to eight weeks. A few orders make the quoted lead times, most don't. Marketing has learned the danger of exact promises and production doesn't know the difference.

Recently, Custom Kitchen Cabinets, Inc., has encountered strong competition from lean producers who quote two-week deliveries on most items. Management would like to improve lead times and continue to grow volumes—but the usual fixes, such as overtime, more operators, the addition of seven high-speed cutting machines, are not working.

It doesn't have to be this way. After 150 years of integrating various manufacturing processes into smooth flows that balance associates with process and materials, lean producers are proving every day that *there is a better way.*

Why Become Lean?

- Customer-centric focus
- Quality product and service
- Increased responsiveness
- Employee empowerment
- Intense competition
- Focus on waste elimination
- Action and results orientation
- Speed

Lean and Beyond

What Custom Cabinets wants to do—make money and grow the business—is what every manufacturer in every industry wants. Despite all marketing and strategic plans, however, Custom is prevented by its own broken and flawed manufacturing processes from accomplishing anything greater than simply making payroll. Bad manufacturing simply won't support good growth and profits.

A map of Custom's main assembly process shows a dangerous mix of batch-and-queue or "push" production. They have a heavy dependence on large, automated equipment and spaghetti-like flows that frequently circle and loop back on each other. With such flows, line-of-sight manufacturing is an impossibility.

Feed the Machine

Further, Custom is drowning in piles of work-in-process and raw material inventory. With so much cash tied up in inventory and storage and handling systems, it will be difficult for Custom to refocus on new products or faster deliveries, especially when the market takes a cyclic downturn. E-commerce will produce a predictable back-room response for Custom.

A Better Way

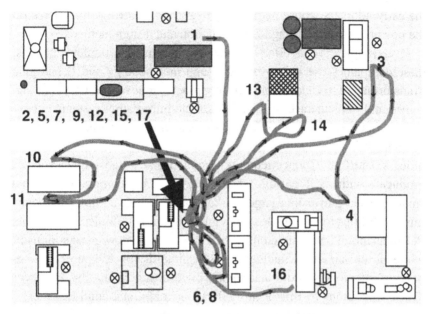

Figure 1-1 The typical operator's work cycle typically involves a lot of wasted motion and extra walking, as shown in this "spaghetti diagram."

Even as customers go online to place their orders, Custom's stressed manufacturing system will run faster and faster to meet nanosecond waves of demand with nineteenth-century methods. It simply won't work.

Batch-and-queue production is inefficient and expensive. Lean manufacturing, especially comprehensive systems like the LeanSigma Transformation, offers producers the opportunity to work smarter with fewer associates, less material, and lower costs. Moving to pull systems such as this—with cellular production, single-piece flow, and repeatable, consistent quality—requires management determination and leadership, as well as visible support.

The Era of Good Manufacturing

Incredible results from elegant, lean manufacturing processes continue to carry the pioneers—like Wiremold, Lantech, Pella, Maytag, Mercedes-Benz, and Vermeer—into new areas of opportunity. These companies are

THE PERFECT ENGINE

the early adopters whose management saw and immediately understood the potential from adopting lean production and design methods.

Over fourteen years after the first North American implementations, these companies use kaizen breakthrough methodology and LeanSigma Transformation to create manufacturing process innovation. They have designed and implemented brilliant but simple information systems to control and track costs, to integrate with the extended enterprise.

Lantech has drawn on the wisdom of kaizen to redesign its design process. Hartford, Connecticut's Wiremold has acquired a dozen smaller producers with its freed-up cash flows. And Pella Corporation, once a small, perfect player in a large market, has transformed its manufacturing capabilities to take the lead in a growing market. In merely six years Pella has more than doubled its sales in a relatively slow growth industry while, at the same time, increasing its profitability by 250 percent, without any infusion of additional capital or resorting to layoffs. Mercedes Truck Operations in Brazil, in the heart of traditional automotive pressures, has demonstrated that manufacturing process excellence is cross-cultural. Maytag is building an innovation machine to compete with Third World labor rates and mass production methods. Vermeer is not only improving its existing manufacturing, it is using LeanSigma concepts to design and develop new machines and new products.

The Challenge

For some CEOs, tackling internal processes such as manufacturing, engineering and marketing may not be hot on their list of favorite activities. Manufacturing is especially unappealing because factories are usually dirty, noisy, even dangerous places. But production is where wealth is created and where customers are captured and retained. With flawed processes, marketing may make the call and close the deal, but without deliveries of high-quality, value-priced product, the customer will be disappointed and sales will dwindle. The challenge is not terribly complex or expensive. Simply put, lean processes can be a big step because of the push systems and heavy assembly line–type flows that are already in place. Moving to a very different and simple system requires parting with tradition and one's comfort zone and going through the pain of

A Better Way

change, and it is the most difficult challenge for management. Essentially, most manufacturers must tear down processes that have worked, although not well, before they can build back up, *the right way.*

Manufacturing in a Changing World

While the 1950s and 1960s may be known as decades of stability and growth, the 1980s and 1990s launched huge shifts for business. Marketing, information systems, and workforce management saw enormous shifts in the philosophy of management, as well as the preferred methods of expanding markets and capturing customers. The possibilities seemed endless if factories could just make enough "stuff."

What began as an exploration into better ways to control electronic processes at Bell Labs in New Jersey blossomed into a whole new focus on basic production processes. Manufacturing became an area of interest, but it was not yet seen as a revenue generator. It was more a cost center line item, or at least a barrier to unlimited market and profit growth.

While manufacturing was just beginning to see the way to more customer responsiveness, MBA classes still taught classic marketing and strategic push planning. The schools churned out executives who well understood how to tempt and push a market until consumers were ready to buy, and willing to wait. Mass media draws like TV ads and even print media were guaranteed investments that would lead consumers, even create a market. How wonderfully predictable and safe this approach proved for so many years to so many companies—giants like Procter and Gamble, the cereal producers, General Motors, and even IBM.

Shifting Consumer Behavior

Manufacturing in a changing market means that the older triggers of newly created consumer demand don't always work, and don't quite buy producers the longer lead times and predictable responses that they had grown accustomed to. Consumer behavior shifts and is as unpredictable as a teenager's fashion sense.

Each new vehicle or entertainment or personal communications de-

THE PERFECT ENGINE

vice is loaded with as many technology devices as the computer industry can cram into products. With the month-long life cycle of most high-tech technology products, it is inevitable that bigger consumer products—like cars and trucks, and even lawn mowers—will undergo life cycle growth and decline that parallels the electronic industries'. Here today, junk tomorrow. There is no way that manufacturers operating at "ordinary" or "traditional" human design and production speeds can "push" into such markets, or even successfully follow and respond to monthly product redesign swings.

New, New, New

Whenever new technologies create new markets, the dominant players' competitive game ups the ante for all the other players. Success draws success, and new levels of furious competition inevitably raise anxiety levels, even for companies that believe they are somewhat protected by patent and trademark laws. Hewlett-Packard's first hand-held calculators were high-end, beautifully designed, but expensive devices. As soon as backwards data entry became passé, however, dozens of competitors rushed in with cheaper devices. Prices dropped from close to one thousand dollars for the first machines to well under fifty dollars in about ten years, with incredible lessons about the power of improved manufacturing methods and costs.

Excellence Is Achievable

We have the answer for many production questions and we think producers already, unknowingly, have the means to become more flexible and responsive, to satisfy a consumer's shifting demands. And fortunately for CEOs, good models of lean production such as the Lantech and Wiremold success stories mentioned above exist now on every continent. There are wonderful lean auto assembly plants in Brazil and Turkey, home appliance manufacturers in Mexico and China, camera producers in Scotland, and dozens of smaller, agile producers at second- and third-tier levels supplying automotive, electronics, and aircraft industries.

A Better Way

Here's an example that highlights the difference between mass production and lean production:

A pen is an assembly of about a dozen metal and plastic components and one subassembly. To build a single red pen using old-fashioned, mass production methods, many big pieces of equipment and a series of human assemblers would be brought together to produce huge batches of shells, cartridges, and clip subassemblies. The entire production process might take two dozen process steps, from cutting, shaping, and painting raw metal to inserting a cartridge supplied by an outside supplier, testing the pen, and wrapping it for shipment with thousands of others.

Typically, if a customer wanted to order one dozen red ink pens and two dozen black ones, he might not expect to receive the completed order for several days or weeks unless it happens to be available from inventory.

The production sequence of operations, following raw material cutting, would include running large batches of individual components through various operations, to eventually meet at final assembly. On the way these component batches move several times back to material storage areas, where they get counted and recounted and wait to be moved again. At every process step, actual batch quantities change due to quality problems—a typical batch-and-queue process.

Actual value adding time, from material cutting, machining, painting, assembling, and packaging, is only a few minutes. But with a batch-and-queue setup, the complete process takes an undetermined amount of time. Nevertheless, eventually, the customer will accumulate the pieces of his original order for one dozen red and two dozen black pens.

Cell-Based Lean Production

Lean production methods, on the other hand, are rooted in simple flow production based on actual demand. Everyone in the operation, and especially the customer, understands what will ship and when, because the entire process is laid out and run to be visible. Machines and operators work to meet the customer's pull signal; cells bring people and material so close together it is possible to change configurations very quickly, and to identify and fix problems equally well.

The pen cell runs with fewer operators and has no automated convey-

THE PERFECT ENGINE

ors or warehousing systems—capital investment and automation are kept appropriately under control. One dozen red and two dozen black pens are produced in minutes, and in sequence, packed and shipped.

> Which approach is more manageable? The flow cell.
> Which approach allows easy expansion? The flow cell
> Which approach produces higher quality with less waste? The flow cell.
> And which approach is the right basis for e-commerce applications to tie production rates to Web-based demand-pull? The flow cell, of course.

As powerful as cell-based production is, it is only a small part of the transformation an enterprise must realize before becoming a perfect engine. Cell-based manufacturing is no more than an exercise in furniture-moving unless you have a change in culture, dedicated leadership, and commitment to customer responsiveness.

This culture of customer responsiveness is a core tenet of the lean philosophy. And we have taken the philosophy one step further, to Lean-Sigma. As a company progresses on the lean journey, and after having reaped the low-hanging fruit, intuitive improvements become difficult. This is when companies require more advanced statistical tools to root out abnormalities in processing systems. For this reason, we have added some of the sophisticated tools of Six Sigma.* When Six Sigma tools are combined with the *lean* sense of urgency—giving birth to LeanSigma—the results take an enterprise from its most basic beginnings to truly advanced levels of improvement. After a decade spent working with hundreds of companies, we knew that a holistic approach was needed to complete a transformation that is both cultural and tactical in nature, enabling an atmosphere where every gain is sustained.

The LeanSigma Transformation

LeanSigma Transformation is how we describe the journey a company takes, from traditional business and manufacturing to one-piece flow and

---

*Six Sigma is a registered trademark of Motorola, Inc.

A Better Way

perfect quality. It is an enterprise-wide process to eliminate waste and create a culture based on continuous improvement and consistent measurements.

The LeanSigma Transformation takes a company from an intuitive level—in which trained operators and engineers can begin to see and fix issues—to the more complex. At the higher level, where issues are not obvious to the casual observer, we use statistical tools to uncover abnormalities. The journey can take an entire enterprise from the first steps to the most advanced—from an initial kaizen creating the first working cell, to deploying LeanSigma black belts to discover the causal effects of raw materials and processing on finished quality. Each level of transformation takes place in a cross-functional team environment, fueled by kaizen breakthroughs, with collaboration and data collection and analysis driving cultural change.

An integrated physical and cultural transformation can yield amazing results. Our experience has shown that companies that have done this with diligence and strong leadership can realize 15 to 20 percent growth in sales and earnings, year after year, without spending any more capital or deploying other assets. And without layoffs. They have learned to exploit the hidden potential of the entire workforce without the restructuring and layoffs that have become so common.

This is a journey that concentrates the energy of an entire enterprise and focuses efforts to serve customers better, faster, with better-quality products and responsiveness, ultimately leading to gains in market share. This is the core goal of a LeanSigma Transformation: profitable growth that serves all the constituents of an enterprise.

Broken down, there are four publics that all producers serve. There is the customer, who wants quality products with unique value quickly and reliably. There are employees, who need a sense of involvement, a feeling of ownership and greater connectivity to the enterprise. There are value chain partners, who want growth of their own company even while they become part of a synchronized, successful value stream. Finally, there are the shareholders. They want sales growth, profit growth, and reliability.

Too many organizations are driven first to please investors, trying to fulfill their needs with every fluctuation of the market. Or some organizations focus on employees, and so they satisfy internal needs and only

THE PERFECT ENGINE

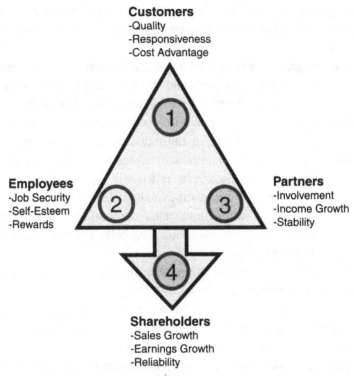

Figure 1-2 The Corporate Challenge is to satisfy the disparate needs of every group served: employees, customers, value chain partners, and shareholders. If the needs of customers, employees, and value chain partners are met, investors will also be pleased.

listen to each other. This focus is inside out, beginning with internal assumptions that can be like mushrooms growing in the dark.

We say that if we meet the customer's needs first—in terms of quality, value, and responsiveness—and we align our employees to help us achieve these goals and help them feel enthusiastic doing it, and we bring along our partners as true partners, then we ensure the long-term profitability and growth of the organization. And we satisfy investors.

Three very profitable and diverse organizations—Maytag, Mercedes, and Pella—tell stories of rejuvenated production strength and flexibility. Hundreds of other companies, and their suppliers, employing collectively hundreds of thousands of associates, have experienced the power of the LeanSigma Transformation:

LeanSigma Transformation Results

Lantech — Started transformation in 1992

Quoted lead time reduction	from 12 weeks down to 2
Annual productivity increase	17%
Annual sales growth	20%
Annual profitability growth	28%

Wiremold — Started transformation in 1991

Quoted lead time reduction	from 28 days down to 2
Annual productivity increase	18%
Annual sales growth	33%
Annual profitability growth	75%

Pella — Started transformation in 1993

Quoted lead time reduction	from 8 weeks down to 1
Annual productivity increase	13%
Annual sales growth	20%
Annual profitability growth	35%

Wiremold increased the worth of the company fifteenfold in nine years, from approximately $50 million in 1991 when its sales were $60 million and profits were slightly more than break-even, to a company that was sold for $77 million in 2000. At Pella, sales have grown two-and-a-half times and the profitability rate has also grown two-and-a-half times as a percent of sales. If Pella were a public company, its valuation would be multiplied five to six times what it was in 1993.

What Pella and Wiremold know is that if a company grows sales without increasing its rate of profitability, net worth growth will be the same as sales growth. However, in both cases, the profitability growth was much higher than the rate of sales growth, resulting in a dramatic growth

THE PERFECT ENGINE

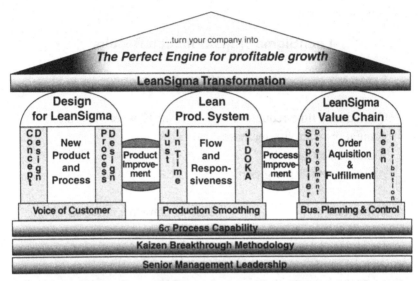

Figure 1-3 In the LeanSigma Transformation model, we illustrate how the modules of Design for LeanSigma, the Lean Production System, and LeanSigma Value Chain are linked by process improvements and supported on a foundation of Six Sigma process capability, kaizen breakthrough methodology, and senior management leadership.

in the net worth. This is the type of value creation that the LeanSigma Transformation is designed to deliver.

Step by Step

After working more than ten years with large multinationals and small companies alike, we have uncovered a few simple truths. Becoming a better company, and a better enterprise, is not a quick or simple process. And nobody *arrives*. Like the engineers of the Toyota Production System who assisted TBM in bringing lean principles to Europe and the Americas, we believe that improvement is continuous. LeanSigma is a process that transforms organizations, but LeanSigma does not bring them to a destination. If you stop, after all, you're out of the game.

There is no end to continuous improvement, but LeanSigma Transformations certainly have a beginning. We have found that setting up companies for success is just as important as giving them the right tools. That's why we discourage a batch-and-queue company from starting its

A Better Way

quality-focused programs, for instance, with the statistical analysis tools of Six Sigma. Those tools are not truly effective until engineers and executives can clearly see and understand their own processes.

Kaizen Breakthrough: Change for the Better

All waste reduction and flow activities are centered on kaizen events. In the beginning kaizen events typically last a week, although the more advanced companies employ shorter, more directed "point kaizens" to attack flow issues. What is important in this case is the team. For each kaizen a cross-functional team of operators, engineers, and business office associates, suppliers, and, sometimes, complete outsiders is assembled. Working together, the team views and attacks the problem from a variety of angles, encouraging creative solutions. Once an idea passes group muster, the team implements immediate change, with full company support. Implementing the Lean Production System is the first step for any company, as teams use kaizens to spur change, shift the company culture, and embrace continuous improvement.

Six Sigma Capability

In the mid-1990s, many companies saw the improvements Jack Welch of GE was trumpeting with Six Sigma and rushed to climb on the bandwagon because in Six Sigma, there was a lot to like. Using methods practiced at Motorola, specially trained experts used statistical analysis tools to uncover nagging quality problems. Companies could find enormous savings in rework and scrap. But there were underlying problems with most Six Sigma programs that few were willing to discuss: the system of black belts and master black belts tended to create elite "masters" who were disconnected from the shop floor and spent, typically, six months on a project.

Being true believers in kaizen methodology—in the sense of urgency and bias for action that make real improvements possible in five days, instead of six months—we studied Six Sigma from several angles before we were ready to implement it. Now we have discovered an elegant synthesis of lean and Six Sigma: LeanSigma.

THE PERFECT ENGINE

As the factory or office becomes a more visual workplace, with each job defined in standard work and organized to fit the rhythm of the customers' demands, the more rigorous tools of a LeanSigma black belt can be used in harmony with jidoka. Black belts attack problems that are more deeply rooted in the systems. Using statistical analysis, LeanSigma black belts and green belts find the root cause of nagging quality issues, and they drive abnormality and variation out of the process. This advanced LeanSigma work is tied to a mentoring system that ensures that each project is supported and that it aligns with business objectives.

From LeanSigma champions at the executive level, to the highly trained black belts they mentor, to the operator-level green belts, each team member knows he is being assisted in this critical work.

Lean Production System

When TBM first began introducing American and European businesses to the continuous improvement principles and techniques of the Toyota Production System in the late 1980s, the focus was on removing wasteful batch-and-queue systems from shop floors, and on creating one-piece flow. This is where all companies begin: reducing inventories and the ugly build-up of work in process that sits between processes.

The next step is to organize factory floors to create simple working cells. Eventually, supervisors walk across assembly areas and see at a glance whether a team is meeting takt time and fulfilling that day's orders.

Using the techniques of just-in-time, work areas are organized to ensure that each operator gets the materials needed to complete the task exactly when he needs it. Just-in-time is a strategy to create one-piece flow, which will eliminate the excess inventory that hides production problems. Jidoka, the second pillar of the Lean Production System, is a system to respond to the abnormalities that just-in-time exposes. Elements of jidoka include visual management techniques, andons or other methods to stop production lines when problems occur, and poka yoke devices. In short, jidoka principles instruct us to pay close attention to problems, to analyze the issues and eliminate roadblocks.

A Better Way

All of these activities must be rooted in a planning system that smoothes out volume fluctuation to make the best use of all resources. If your company sells an item on a daily basis, we say, you should build that item every day.

In no time, these lean activities branch out from factory floors and into the business office where teams attack wasted motion in order fulfillment, customer service—even the accounting department. At two of the best lean companies, for example, Wiremold and Lantech, accounting executives now produce real-time financial reports, not two weeks after the period's close, but on the same day. They have also found innovative ways to reduce the amount of time spent chasing daily invoices and transactions.

From the business office, the Lean Production System moves upstream to suppliers and downstream to distributors. Each company in the value stream becomes leaner, more profitable, and more capable of the flexibility demanded by today's markets. We have seen that discrete companies linked by common vision and goals can become partners instead of adversaries.

Design for LeanSigma

One critical step in the transformation process is applying LeanSigma principles to the biggest, most important investment any company makes: product development and launch. This step involves knocking down initial capital investments and carefully building quality into every product by reducing variation in the process. With Design for LeanSigma, teams concentrate on three vital areas of product development, ensuring that poor assumptions and bad process are rooted out before they are hard-wired into a product:

- *Concept Development,* which is the translation of the voice of the customer into product specifications and conceptual drawings or models. Key tools might include a House of Quality, which helps formulate customer requirements, product specifications, and competitive benchmarks into a product concept.

THE PERFECT ENGINE

- *Design for Manufacture and Assembly* focuses on minimizing the number of parts required for design, to ensure quality and make sure it can be produced within Lean Production System principles.
- *Process Development* is preparation for production: creating assembly lines, material flows, and layout that are simple, functional, and ensure ongoing flexibility. Companies that are leveraging growth into new plants also use the principles and tools that apply here.

The best lean companies are using newfound strengths in Design for LeanSigma to aggressively compete in their own markets with product innovations and flexibility that their competitors are not prepared to meet.

LeanSigma Value Chain

Finally, the LeanSigma Value Chain reflects TBM's emphasis on integrating and aligning all elements of an enterprise to create a synchronized system that rests on the foundation of business planning and control tools. Instead of focusing exclusively on the activities of one shop floor, one business office, and one company, our methodology takes into consideration the entire value chain—from suppliers to the distribution network and customers. Using strategies in supplier development, we assist companies as they transform suppliers into reliable partners, while teams improve logistics and distribution channels. In a time when electronic linkage can pull product on demand directly from suppliers, producers, and distributors to consumers, with little time and few steps in between, each entity in an enterprise must be connected to each other. Working in the Value Chain, we create line of sight across companies.

Evidence of the LeanSigma Transformation is visible now in Brazil, in Mexico and China, France and England, Iowa and Connecticut. But nowhere, perhaps, has it been more dramatic and quick than in the hill country of Cleveland, Tennessee, at Maytag's center for cooking products at the foot of the Great Smoky Mountains.

A Better Way

Maytag's Industrial Evolution

The heart of Maytag, headquartered in Newton, Iowa, is innovation. The Cleveland, Tennessee, plant is the launch site of many new products for Maytag, as well as its first steps in a long series of successful kaizen activities for manufacturing transformation.

Harding Gamble, age seventy-four, a Maytag veteran who started in the shop at age sixteen in 1939, describes Maytag's transformation: "I have never in my life seen anything changed as much as I have in this operation in the last few years. Compared to the way we used to do things, and the way we do them now, well, there's just no comparison."

Harding's observations spring from the LeanSigma initiative in Cleveland, which combines lean manufacturing and Six Sigma principles with intelligent innovation.

Harding witnessed the change when Maytag Cleveland Cooking Products converted its recreational-vehicle cooktop and range assembly

Figure 1-4 The Maytag Repairman—also known as Ol' Lonely—is symbolic of Maytag's high quality. Innovation is the heart of the company.

19

THE PERFECT ENGINE

areas to lean lines. Work first started in 1997 and has expanded into all areas of the Cleveland plants. The objectives were rapid improvement in productivity and quality and elimination of waste.

The Cleveland plant is making history, proving that new products and processes offer opportunities to update its more traditional product lines. Physical plant challenges—nineteenth-century buildings on hilly terrain, connected by driveways and parking lots—complicate the layout and process flow with varying degrees of difficulty.

While the physical plant may remain a challenge, however, the people are Maytag's greater promise. It's not the machinery or the automation, and certainly not the buildings that appreciate. The physical assets of any business inevitably depreciate, but the people appreciate constantly. When companies invest in people, give them the right tools and training, and put them in an environment in which they can improve continuously, the entire company benefits from a constantly appreciating asset. Investment in human capital results in more positive employees—adding value to the company and shareholders alike and, most important, giving the customers better quality and delivery. Maytag's dedication to investing in its workforce has caused the company to expand its LeanSigma programs. Hourly associates completing forty hours of training earn three hours of college credit toward a two-year Associate Degree at the local community college. Teaching combines both classroom and shop floor activities.

For workers this is a serious investment in keeping their livelihood in the rolling hills of southeast Tennessee, as well as improving their quality of work life with better job design and ergonomics. It is also at the heart of Maytag's strategy to compete with GE and Whirlpool initiatives. Associates have removed inventory and improved housekeeping and quality as well. Perhaps Maytag's greatest contribution to the application of lean manufacturing in new areas is its intensive implementation of Design for LeanSigma.

The Power of Simulation: Design for LeanSigma

In August 1998, design teams assembled to create an efficient production process well in advance of typical manufacturing engineering cycles. Associates barricaded themselves in the simulation lab to design, develop, and even simulate manufacturing a new cooking product scheduled for public debut before the Maytag Gemini made its appearance in 1999.

Working shoulder to shoulder, eighteen designers, engineers, quality technicians, and upper management created a mock assembly line on which simulated models of the new cooking product were actually built. Next, team members assembled the pilot unit.

Design for LeanSigma Rules

1. Creativity before capital
2. Quick and crude versus slow and elegant
3. Simple, inexpensive, and dedicated equipment
4. Develop a minimum of seven alternatives for each issue
5. Select the three best designs and create lifesize mockups
6. Combine the best features of three into one
7. Simulate the best design and the process before committing to hard design or process investment

Ramin Zarrabi, Kaizen Promotion Office head, said, "We are making discoveries now that normally aren't made until far into the manufacturing future of a new product—this process brings all LeanSigma principles into the initial stages of a new product. Unlike traditional kaizen, this process doesn't wait to fix what's broken in an existing manufacturing process. Rather, it starts from scratch."

Dan Sullivan, a TBM managing director, sees Design for LeanSigma as the way for an organization to make changes in its process. "We do it from the get-go, from the very start, to minimize the trauma of new product launches."

Gregg Greulich, senior director of R&D, is just as enthused about Design for LeanSigma. "This technology is filling a huge gap in the whole preproduction process. I'm excited about this from the design standpoint because it allows engineers to design a product for assembly and not

THE PERFECT ENGINE

with the idea of designers drawing it up and hoping shop floor operators can build it."

Tom Briatico, Cleveland vice president and general manager, believes "the most powerful feature of this development tool is driving quality at the source and mistake-proofing the assembly process."

Maytag Results

> Set your goals and targets high because you are competing against the entire world.

Throughout its Cleveland operations, kaizen leaders have racked up impressive results. In the West Plant, team members reduced cycle time and learned valuable lessons in waste reduction as they right-sized equipment and rebalanced lines. Shop-floor operators, rather than engineers, redesigned work stations and redistributed job assignments themselves.

In fabrication, a line of five 250-ton presses achieved 56 percent reduction in setup time using standard setup procedures, visual displays, and performance boards to maintain standard work. Their application of LeanSigma methods to a highly complicated porcelain process helped to substantially improve first-pass yield and identify what variables cause defects. Using the one-week kaizen structure and applying Six Sigma tools of process maps, FMEA and DOE, defects were reduced by 60 percent. This approach also increased throughput.

In the year after TBM started working at Maytag associate teams conducted over 124 kaizen events, and more than 1,400 employees participated. The kaizen events yielded improvement in the unit output per employee from 25 percent to 35 percent; floor space was reduced by 83,000 square feet, or 27 percent in the affected areas. Line inventory was reduced by 50 percent to 60 percent. Team members also produced between three and five safety improvements per event. Repair rates declined by 75 percent. Kaizen impact also avoided $7 million in capital expenditures, Maytag estimated, including $5 million related to lean versus traditional approaches (e.g., conveyors), $1 million in equipment for new products, and another $1 million for capacity.

A Better Way

Gemini was the first new product to apply lean concepts to assembly. Gemini results included:

- 40% reduction in service call rates
- 20% reduction in final assembly labor
- 50% reduction in line length
- 50% reduction in repair rates

Although Maytag Cleveland associates realize they have much to do and many more areas to transform, the early lessons from going lean are powerful testimony to their pioneering work. Tom Briatico has come to believe that lean implementation is less about physical change and more a change of mindset as we think about how we work.

Ten LeanSigma Lessons Learned at Cleveland

- Real change does not take place in the classroom
- Start small and learn from mistakes
- Respect shop floor employees (management's customer)
- Learn on the shop floor (top management)
- Maintain an obsession for defect and waste elimination
- Design and build small and inexpensive equipment
- Have an obsession for maintaining standard work
- Design methods so operators cannot make a mistake, or pass defects on to the next operation
- Create a model line as a learning laboratory where the transformation should be deep vs. wide
- Be prepared to uncover many rocks—treat these discoveries as golden opportunities to make the process robust and free of abnormalities.

E-Supply Chain Strategy

Giving customers what they want means also meeting demand for high variety and customization on demand. The Build-to-Order Project launched

THE PERFECT ENGINE

in late February 2000 was designed to pick up where simple Web-based order taking left off. We could all see that there was a big gap between Web-enabled organizations that take orders, create new designs, manage their supply chains, and ship via Web connections at nanosecond speed, and the "fronts"—businesses that use e-commerce for order taking with no back-room support. Big warehouse networks and premium shipping arrangements, like the ones maintained by Amazon.com, are an expensive substitute for truly e-enabled production networks.

Ford and GM think they have pioneered it, and hope they are moving toward the 3-Day Car envisioned at the Tokyo Auto show. In the appliance world, Maytag wants to do Web-based business the right way.

Maytag's Ramin Zarrabi and Sam Swoyer of TBM, build-to-order project leaders, recognize that their group's vision is an aggressive one: customer orders on day one, product ships on day five. Charged with implementing four or five examples in twelve to eighteen months, the proposed systems will build and showcase Maytag and its partners' capabilities, as well as serve as an enterprise-wide beta experience.

Five pilot project possibilities are:

1. Collaboration between Maytag Cleveland and Fleetwood, a California recreational vehicle producer. As each small cooktop/oven unit was installed in a California RV, the assembly line in Tennessee would know immediately via electronic kanban, triggering production of another unit.
2. A new high-end refrigeration product, code-named Alaska, that is built to customer order at low volumes, with considerable internal variety.
3. A high-end outdoor grill destined for home users.
4. Commercial cooking units fitted to upscale hotels.
5. A Hoover commercial product.

Art Learmonth, Maytag vice president of manufacturing and engineering, tends to be a little conservative, in his own words. Learmonth believes that *process*—clear, simple, degreased good process—takes precedence over big software solutions. Naturally, Learmonth is anxious to test the robustness of Maytag lean manufacturing muscles by leveling production and rendering it more responsive and flexible. Along the way,

A Better Way

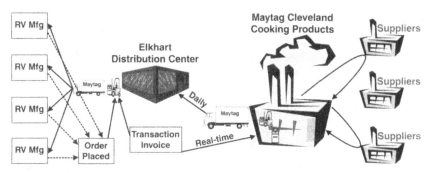

Figure 1-5 A value chain map, showing the collaborative efforts of Maytag and Fleetwood RV toward an e-enabled replenishment system. The plants may be on opposite sides of the country, but information and product flow is immediate.

Learmonth believes that by executing a perfect e-enabled operation, Maytag will have proved once and for all that this business does not need pockets or trailer trucks of inventory to get by.

Mercedes-Benz do Brasil

In 1953 Mercedes-Benz do Brasil, a division of DaimlerChrysler headquartered in São Paulo, began an expansion of operations to produce trucks, buses, and passenger cars to support domestic and export orders. Thirty-six years later, three plants produce over fifty thousand vehicles.

Employing more than twelve thousand associates in Brazil, Mercedes captured almost half of the domestic luxury car market, 70 percent of buses, 14.1 percent of light trucks, and 33.9 percent of trucks. Total corporate capital investment stands at $308 million, U.S.

Clearly, investment in Brazilian operations has reached levels of strategic global importance for Mercedes. And so have the systems and training the company has poured into its three model plants.

The success story is quite young, as the new Brazilian economy has just begun to settle into tier-one levels of productivity and efficiency, in less than ten years.

The Kaizen Breakthrough

In 1991, the challenge for most Brazil-based manufacturing companies was simple survival. Although the Brazilian government legislated a change

THE PERFECT ENGINE

from a closed to an open economy, businesses continued to be buffeted by continuous inflation and strong external competition. High-cost inventory and production practices had to change for companies to stay in the market.

Mercedes' Karsten Weingarten had heard of the power of kaizen, and a tour of WABCO, a division of American Standard, persuaded him. WABCO's first kaizen reduced setup time on a double-spindle boring machine from four hours to twenty minutes. The time to set up a numerically controlled lathe was cut from one hour to four minutes. Overall, assembly line productivity—another desperate need for so many Brazilian plants—increased significantly as work-in-process inventory dropped by 82 percent.

Mercedes' successful introduction of kaizen and LeanSigma Transformation is notable for its iterative, comprehensive approach. Starting with sixteen kaizen events in 1994, the operation has steadily increased its commitment and kaizen expertise to a rate of about seventy events per month. Improvement opportunities cover a full range of plant operations and processes, from rear axle assembly to gear and differential case manufacturing cells.

Transformation Results

One thousand five hundred and twenty-seven kaizen events at Mercedes-Benz in Brazil produced these results:

Area	-43.2%
Productivity	+30.0%
Inventory	-46.0%
Setup time	-64.0%
Lead time	-92.0%

"Under One Flag . . . the Best Axle Manufacturer in the World"

The Brazilian operations are a diverse global mix of workers from different regions, religions, and races, which is another reason for associates' pride in their achievements "under one flag." They have created a faster, cleaner approach to vehicle production, rooted in Henry Ford's vision of integrated production. With smaller, dedicated lines, the pull process reaches all the way back to first-tier suppliers. Production smoothing of the assembly lines and its feeder operations are a model of lean processes. Computer systems found to be a barrier to lean flows were shut down so that the vehicle assembly line schedules could be sequenced for three days, giving materials pros three days to execute parts plans.

Weingarten's goals of agility, quality, and profitability, incorporated in a plan called Factory 2000, hinged on worker involvement. Sixteen full-time kaizen experts trained by TBM work in-house to conduct and plan LeanSigma work.

What started out as a seemingly new, alternative concept in manufacturing philosophy has mushroomed into a dynamic grassroots revolution which may shortly become the norm. After this impressive transformation, LeanSigma concepts are now being applied at DaimlerChrysler operations in Argentina, Mexico, the United States, and Turkey with the help of TBM consultants.

Pella

Pella Corporation, headquartered in Pella, Iowa, is a producer of premium windows and doors. In 1999 Pella was the second-largest window and door manufacturer in North America, with seven U.S. plants and one in the Netherlands. Employment totaled five thousand directly employed by Pella and one thousand people employed by eighty one-step distributors in the United States and Canada. Expansion into new market segments and distribution channels, including one thousand Home Center locations and various lumberyards, promised unlimited growth in market share and profits.

But before the company began its kaizen journey, Pella's world was somewhat limited by operations' performance. Although the Pella stamp on windows guaranteed quality and product innovation, in 1985 the marketplace was indicating that Pella products were expensive. Long lead times and unreliable delivery performance made it hard to do business with the company and its distributors.

Further, there was a certain complacency about market growth that

THE PERFECT ENGINE

masked changing business conditions. In 1987, when sales took a dip and new product introductions taxed Pella's traditional production and distribution resources, it was clear that Pella needed a new, reliable, consistent process.

In December 1992 three Pella executives from operations, engineering, and finance attended a public Shop Floor Kaizen Breakthrough event conducted by TBM, hosted by Carrier Air Conditioning in Arkansas. The Pella executives knew the climate was ready for big changes and saw the possibilities in the kaizen breakthrough. One month later, in January 1993, Mel Haught, Gene de Boef, and Herb Lienenbrugger arrived on the shop floor wearing jeans, ready to begin. Within the week, Haught became Pella's internal kaizen champion. Seven years and thousands of kaizens later, Pella has become a model continuous improvement pioneer.

Results

Pella associates have experienced the power of kaizen in all areas of the operation, from engineering and production down into the supply base. Kaizen work started on the shop floor, but by 1994 Pella had incorporated white-collar improvement in its Business Process Improvement Program, championed by the senior vice president of finance. Design for LeanSigma teams headquartered in Pella's spacious Product and Production Preparation lab created brilliant and innovative solutions to process flow and shop floor challenges.

In 1994 Pella began a long series of design events with Gene de Boef, vice president of engineering, as corporate champion. Because 85 to 90 percent of the cost of new product is locked in at the time of product and process design, Pella was determined to design right the first time. Team members focused on improving ergonomics, removing complexity, using creativity before capital, and accelerating the concept-to-market cycle.

Results of their labor appeared early, including overall reduction in development time by over 50 percent. Further, team members believe that designing processes to support only next year's volumes will ensure profitability at startup, not three years down the road.

A Better Way

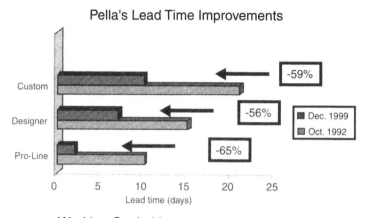

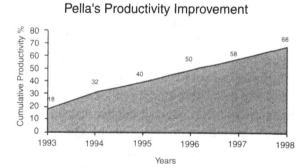

Figure 1-6 A sampling of Pella's improvement charts shows lead time dropping by 56 to 65 percent, work-in-process inventory turns going from 12 to 50, and an average annual productivity improvement of 11.3 percent since 1992.

THE PERFECT ENGINE

One of Pella's most outstanding and visible contributions to manufacturing reform is its innovative approach to capital expenditures, the machines and facilities that support process. Not surprisingly, when Pella kaizen associates advanced through various departments, removing big equipment monuments and replacing the behemoths with smaller, smarter solutions, the writeoff of fixed assets ballooned.

Lessons Learned—Business Process Kaizen

Most of Pella's "Customer Hassle Factor" and untimeliness was administrative, not operations related. This finding was an eye-opener. Planners discovered that about one-third, or 28 percent, of the seven-day total lead time could be attributed to manufacturing. Eighteen days, or 72 percent, was administrative. Because of this performance gap, Pella launched a Business Process Improvement initiative in 1994, under the oversight of the senior vice president of finance. Teams attacked paperwork flows from product concept through launch, even including review of the new employee hiring process and monthly financial closing time.

Benefits Not Flowing to End Customers

With such incredible results, Pella executives were surprised to learn that despite cuts in processing time all over the plant and the office, benefits were still not flowing through to end customers. It was as if all the good results were loaded into a narrow-ended funnel and one by one, bit by bit, small advantages leaked out to consumers.

Nineteen ninety-five saw the beginning of a concerted effort to open up the valve. Pella associates conducted the first kaizen events at independent distributors—an effort that created the Office of Distributor Process Improvement, headed by the senior vice president of marketing and sales. Again, high-level executive support stood behind every associate's effort.

Another lesson kaizen team members learned at Pella was that kaizen efforts were not coordinated—team members were fixing pieces of many, many processes instead of all the pieces of a single process. As

A Better Way

powerful as individual projects were, their impact was limited to specific flows. Team members wondered if more integration would get kaizen work done faster, with less pain. In the next two years, teams began a program to identify core and supporting processes, with the goal of understanding and managing the whole company by process.

Pella's wonderfully creative and quietly dedicated workforce is key to how the company continues to rack up success after success. Seven best practices continue to keep the spirit alive:

1. Dedicated resources, reporting to senior management, including the development of an Advisory Committee, and the Kaizen Promotion Office
2. Management carefully monitors and adjusts for possible employee kaizen burnout
3. Kaizen rolls into the MBO (management by objectives) process
4. Strategic planning process includes kaizen programs
5. Twenty-five percent of profits are shared with employees
6. CEO's message: When in doubt, be bold
7. Celebrate the wins!

President and CEO Gary Christensen says, "Fundamentally, kaizen works and kaizen is one set of disciplines that unleashes human energy and creativity. It's noteworthy that the principles of kaizen run headlong into the principles of the way business management had been executed in the sixties, seventies, and eighties—the command-and-control John Wayne version in which one cowboy rides in and makes all the difference. But that symbolism is not nearly as powerful as what organizations and people working in concert can do."

Christensen has a visceral knowledge of the challenge this presents to his corporate peers. He recalls that after one kaizen event, co-author Sharma asked him whether he was pleased with their impact at Pella. Christensen assured Sharma that he was and expressed deep thanks, but said, "But I do have one bone to pick with you.

"When I was a young man and was making presentations, I had great passion, but I learned to ask permission to do certain things. Now I find myself turned around—the decisions get made by the teams during the kaizen and I have to come to a kaizen presentation to learn about them."

The reversal, says Christensen, "speaks to the power of kaizen, to the difference in the management environments, and why it takes somebody who absolutely knows it works to be a change agent, to help the company. They cannot, however, do it just for the company. I think the reason kaizen is having such a positive impact in Pella is that we know that kaizen also makes a difference in the lives of our employees."

Global Success Stories

Unipart

At England's Unipart, which manages aftermarket and spare parts for Jaguar and Rover, workers proved that a LeanSigma Transformation works in distribution and logistics operations as well as it does in the basic production. By improving layout and reducing space, teams were able to integrate two other businesses, Hewlett-Packard and Honeywell parts distribution, in the same facility.

Associates redesigned the process of ordering parts, stocking, picking, packaging, and then delivering orders to dealers. To do this they improved relationships with suppliers, and developed twenty-four-hour emergency delivery systems. Mike Herr, TBM Europe's managing director, believes the Unipart experience underlines a basic lean principle: Don't lay off people; use the gains to grow the business.

Black & Decker

Another British operation, Black & Decker in Spennymore, County Durham, in northeast England, is a high-volume producer of power tools like drills and lawn mowers. The company saved one million pounds in labor costs in two and a half years of kaizen activities. Black & Decker drastically reduced space requirements, enough to close its plant in Italy and bring production back to England. All these improvements were accomplished without additional bricks and mortar. New production methods have improved turns to forty per year, compared to the high teens in prekaizen years. Black & Decker was also pleased to be able to reduce temporary labor.

Polaroid, Hill-Rom, Tofas

Polaroid's Scotland facility, faced with a corporate threat to outsource production to China, has survived the threat and developed a strong business base supporting innovative production, three years after its LeanSigma Transformation began. Hill-Rom France, and Turkey's Tofas, an auto plant whose supply base was devastated by earthquakes, all point to the power of kaizen and workforce engagement.

Think about it: every minute, one perfect car. A near magical assemblage of 6000 parts from 300 suppliers touched by dozens of human and robotic hands, slips off the line at hundreds of auto plants around the world every minute. LeanSigma Transformation, a product of the individual contributions of Henry Ford, Toyota's Taiichi Ohno, Shigeo Shingo, Dorian Shainen, Robert Shewhart, and thousands of empowered production associates, has spawned a revolution in quality and production process. Lean principles have been implemented across all industries, at all levels—from first-, second-, and third-tier automotive suppliers, to aircraft manufacturers, plastics, electronics, and even financial services. The combination of lean manufacturing principles, Design for LeanSigma, the Lean Production System, LeanSigma Value Chain, and an eye to the next generation, Distributive Manufacturing, is an exciting, redrawn approach to sustained growth in sales and earnings.

If you want to transform your enterprise to grow faster with fewer resources, better quality, more responsiveness, you must do it different. And better.

CHAPTER 2
WHACK!
What Doesn't Work and Why

Things May Not Always Be What They Seem

Operators working at a furious pace next to crates and shelves and boxes of dusty, silent parts, waiting, waiting . . .

Racks of incomplete printed circuit boards careening from one building to another . . .

Managers who bravely tackled MRP, then MRPII, JIT, Lean Manufacturing, and maybe even a little ERP—what's next?

E.com Websites and order processing backed by nineteenth-century warehousing and stockpiled inventories . . . Click and wait, wait and click . . .

Employees standing ten inches away from each other whose sole human contact happens at break time, and lunch, and when the 3:00 P.M. bell rings . . .

Ten-ton machining centers whose ingenious computer controls and tooling require days of preparation to run one simple cutout.

New product designs whose processing takes more man hours, more fixtures, more fixes, and more steps than their predecessor's.

Herds of self-directed work teams lurching from benchmarking industrial tourism junkets, to explorations of work flows, to setup time reduction projections, to team meetings, to end-of-the-latest-

WHACK! What Doesn't Work and Why

initiative celebrations, followed by repeated death-through-boredom team meetings . . .

Gorgeous products, filled with promise, whose profitability becomes questionable too soon—the market is there, but the process ain't.

Customers who scream for cordless black nail guns in the morning and industrial orange power drills by lunchtime.

Boatloads and truckloads and 747 loads of "outsourced, offshore" components, constantly in transit, circling, waiting, clearing, packing, trucking, unpacking, touched by hundreds of hands before even one customer reaches for the box . . .

It's About Learning to See

When manufacturing was a back-room black box fronted by a glamorous advertising/marketing machine, as long as the profit and loss statement showed operation's line items run in the black rather than in the red, how the process actually worked "back there" was not important. The idea was that if we bought the materials reasonably well, brought in a sufficient number of low-paid bench workers or machine-minders and periodically opened the big metal doors, piles of products would stream out, eventually to be traded for cash, receivables, and new product funding.

A Lesson in Observation

The director of total quality for a Big Three supplier, Allan Crames, experienced "learning to see" in a very direct way. It seems that his company had run through a decade of various quality improvement initiatives. The consulting and overtime dollars had piled up, but the hoped-for results just never came in. Still, customers continued to demand Six Sigma product quality and delivery performance.

Crames was raised on statistical quality control and traditional manufacturing processes. Still, when a member of his company's latest quality improvement team invited him on board, he did not hesitate. His first assignment, however, came as a surprise: Handed a yellow legal pad and

THE PERFECT ENGINE

pencil, he was sent out to the floor to observe—simply and quietly—for one entire shift.

This executive admits that he was indeed caught somewhat off-guard—a pep rally or some statistical analysis he might have expected—but sitting on a wooden stool at the end of one of the parts lines was the last item on today's to-do list.

Nevertheless, Crames took a walk and after an hour out on the floor, with a few notes in hand, he returned to the war room, armed with pungent observations and brilliant recommendations. He was met, however, by the sensei: "What are you doing back here? It is not time," and summarily sent back to his stool.

Another hour passed and Crames began to see things differently. A pileup at the end of his line seemed to be accumulating in one small box—no one from the downstream operation seemed to know his order was ready, and it looked like some of the small plastic components for this shipment had fallen off the shelf to a bin on the floor.

On the other end of Crames's observation post, operators were struggling with a small drill press—tools were missing, adjustments had to be made—until finally they gave up and wandered into another work area.

Further upstream, an assembly line expeditor dropped in, chatty and friendly, but waving his hourly shortage list in the foreman's face. Word of another hot order passed up and down the line as various operators searched to identify the missing components, dropping their work to fill the hot order.

And so Allan Crames's morning passed, filled with observations of process issues—missing parts, problems with machines, a little bit of breakage, and interrupted flows. Was this what quality and robust process were all about?

Crames decided to abandon his urge to quantify the movement of components, and instead sketch the process as he observed it in the afternoon. He had an idea that if he could simply observe the day of a single operator, or possibly follow the path of one plastic dashboard assembly, he would uncover even more process surprises.

He was not disappointed. Coming back after lunch break, the operator he chose to quietly observe was disappointed to find that not only had his work area been rearranged—apparently maintenance had been by to at-

WHACK! What Doesn't Work and Why

tempt a fix—but the press had disappeared! A few walkabouts and inquiries located the now disassembled machine out behind the parts racks. Looked like nothing was going to be happening with this piece of equipment this afternoon.

Sometimes the simplest approach is the best, and Crames decided it was time to hit the floor one more time. Armed with a fresh yellow legal pad, he left his stool and followed the bill of material routing of a typical dashboard component, all the way back to raw material. It was a long process, visiting each operation and storage and inspection spot along the way. By the time Crames had a formed plastic frame in view, the part was well into a seven-mile journey through the plant that ended several days later at his original observation post.

Strange, but by following a single part, rather than anonymous bins of semicompleted components, Crames was able to rough out a sketch that started to take on the look of an unfinished interstate highway system, a few solid lines interrupted by dotted lines and lines that went nowhere. In effect, what Crames had so painfully observed and attempted to map was a flawed process flow whose structure had been dictated by a twenty-year-old traditional manufacturing push process methodology, complete with order release tickets, red reject tags, roof-high racks of parts storage, and all the other signs of an MRP environment gone awry.

Bad Process Dead Giveaways

Trust your senses to point to evidence of bad processes—sight, hearing, touch, even smell are still the best way to understand how a process is working.

Sight:

Are the operators working at a very fast pace? Then do they slow down and wait? Do expeditors and supervisors periodically converge on operators to solve problems?

(Continued)

THE PERFECT ENGINE

Can you track the progress of a single part from beginning to end of process by line of sight?

Is there any rhythm to the work of the operators?

If you notice a bolt or piece of scrap paper on the floor, will someone rush to pick it up and drop it in a nearby receptacle?

Are floors painted and marked with safety lines?

Are there high storage racks, warehouses, rented tractor trailers in the parking lot?

Smell:

With your eyes closed, can you tell by the odors in a plant what operation you were in—foundry, a paint shop, and so forth?

Hearing:

Punch presses, when they are in operation, run to a regular rhythm. Do the machines squeak, grind, and cry for help?

Touch:

Are machines covered with dust? Are the floors slippery to touch or dangerous to walk on?

When you run a finger along the top of stacked boxes, do they flunk the white glove test?

A Change in Perspective

For CEOs, until recently, managing a business typically did not require much insight or concern about managing the back room, or the factory downstairs, or out back, or over in China. Manufacturing was a messy process best left to the engineers, the facilities people, and the people minders—personnel and foremen, or the union, and shop supervisors. Running a good operation assumed that technical issues got solved, and meeting the marketplace demand presupposed that all the resources—machines, people, material movement tracking, and expediting systems—would somehow be put in place.

WHACK! What Doesn't Work and Why

White Socks Take Center Stage

But, as so many companies with a narrowed vision of the role of manufacturing in their overall success have learned, it is a mistake to underestimate the role manufacturing operations play in profitability and time to market. Manufacturing has moved out of the periphery of management vision right square into the CEO's line of sight. It has become clear to every Detroit and transplant auto assembler, and thousands and thousands of tier-one, tier-two, and tier-three suppliers, that the few quality defects just one of them eliminated among six thousand bill of material parts can truly persuade a consumer to switch from the Accord to the Camry, or from Ford's Explorer to the Toyota 4Runner.

When quality and features are assumed—they become tickets to the newly empowered consumer-driven marketplace—cost becomes a big differentiator, as does time. Time is a function of cost inside manufacturing operations, and CEOs who ignore the internal clock of their production systems risk unstable profit margins. Ugly surprises to the board don't make for fat executive compensation packages, coming or going.

Clearly, the rules of business have put manufacturing operations in the center of the run for profits and market growth. The key strategic position, however, is not always clear or well understood when ordinary issues—shortages, bad quality, workforce issues—exhaust management attention over more strategic questions, such as when to build that stamping plant in Mexico, or which dot-com can be put to work harvesting profits.

And for senior executives who may not be as prepared as engineers or operations pros for seeing and understanding the outlines of a good process, or the danger signs of an out-of-control operation, the obvious answer, to hand *it*—running those messy operations—off to hirelings has limited usefulness. Because a CEO or a president and a board of directors are responsible for creating wealth, and they are tasked with building a company's future, senior management has no choice but to dig in and discover for themselves the best way to fill the marketplace with gloriously successful, elegantly designed and marketed company progeny. In effect, any CEO who wants to understand and leverage his manufacturing operations must first learn to see.

THE PERFECT ENGINE

Learning to See with Different Eyes

We know that by improving the manufacturing process, we will improve quality.

The Toyota Production System, Honda BP, and various individual solutions—JIT, for example, and Motorola's creation of its comprehensive Six Sigma program—offered a wealth of parallel improvement approaches. Some methodologies laid out a detailed road map of recommended sequences of improvement priorities. Companies were aware of the possibilities and benchmarked the best practice models, but few of them swallowed the whole system approach.

Instead, they bit off small chunks—a little bit of Toyota Production System, some Six Sigma quality disciplines, or supplier development techniques from Honda. As a starter, many North American producers quickly learned that just by perfecting setup time reduction methods, they took endless waiting hours out of cycle times, as they produced in smaller daily or hourly batches and established a new awareness of the value of flexibility in factory operations. They were learning to see.

No Cure-Alls in Sight

Still, progress throughout the seventies, eighties, and early nineties was erratic, as improvement zealots lurched from one obsessive focus on operations to another, their vision distorted by whatever lenses management acquired. Baldrige, Deming, ISO certification, inventory reduction, operational excellence, even self-directed work teams were launched without a comprehensive, systematic objective in mind.

Push-Button Manufacturing Prowess

It seemed that the business of running a manufacturing operation was still perceived as random button pushing—for problems with the workforce, push the money button; for problems with shipments, push the overtime button; for problems with quality, push the inspection button.

WHACK! What Doesn't Work and Why

Things Really Aren't Always What They Seem

Along the way, buckets of dollars poured into various solutions: acres of bricks and mortar, lights-out factories, offshore production, thousands of miles of conveyorized lines, and complex software tools—bar coding, and sensors, automated warehousing schemes, and computer-controlled parts inspection and flows. If we did nothing else during the sixties boom, and the seventies and eighties upward climbs, we spent money on plants and capital equipment. It was an industrial engineer's dream.

Despite the enthusiasm and hope attached to each of these marvelously engineered solutions, the success they promised never quite materialized. And unfortunately, as interest inevitably waned, the new plant's or new line's skeletal remains, fragmented outlines of odd pieces of new solutions, lingered on the landscape. Unseeing managers and associates still regularly uproot and trip over the bits and pieces of these early attempts at process improvement.

General Motors

A couple years ago TBM began an engagement with a division of GM. The agreement was that the project could only go forward with upper-management leadership and support, and work would be focused at two plants, one making shock absorbers and the other making brake assemblies.

Initial kaizen activities showed promise: Despite 100 percent union membership, the workforce was excited and encouraged. The project team was able to quickly maneuver around the usual work policy challenges; management and supervision began to see their operations in a new way—with clean, clear process, and uninterrupted flows of near-perfect product. It was an operations manager's dream.

Good News Travels with the Speed of Modems

The word got out—the brake plant was moving equipment around, people were switching jobs, all hell was about to break loose. Truth travels faster than rumors, however, and within weeks, someone from headquar-

THE PERFECT ENGINE

ters—"I'm from corporate and I'm here to help"—pulled up, looking for improvement opportunities, no doubt.

The Problem Is with Machine Utilization Rates?

The corporate guy had a puzzling message to deliver. There was a bit of internal competition going on with the outside experts, and the insiders were determined to win. "We've looked at both approaches, and corporate is convinced we need to go with *my* approach. See, the problem is that this plant is running at about 45 to 50 percent equipment utilization, and you know the best Japanese plants do 85 percent, so if we approach the first step improvement objective as achieving 60 percent utilization—all we need to do is to *cut the takt time* (takt time is a term used by Toyota to define a time element that equals the actual demand rate) to 60 percent. We will make huge progress. Move over, it's been decided."

Not all creative solutions, however, are grounded in the science of manufacturing, and this wild-eyed idea ranked with the most creative.

"You must be kidding," co-author Sharma responded. "You think takt time is linked to equipment utilization. . . . You say the machines now work 45 percent of the time and 55 percent of the time the machine is down—doing nothing—so your solution is to *cut takt time,* and put 40 percent more people in the process, more people than this operation needs. How is that progress? Now, when the machine goes down even more people will be doing nothing!"

Corporate Manager of Takt Time

Nevertheless, the convoluted logic continued to spin a wider web, pulling organization structure and job descriptions into the corporate expert's vision of shop floor improvement. The next step, appointment of a corporate manager of takt time, spelled the beginning of the end for this improvement initiative.

Rather than attacking hard shop floor problems—the root cause of machine breakdowns, scrap, bad flows, and flawed ergonomics—the company was doomed to pretty up a bad process with cosmetics, words that conveyed the improvement message, unsupported by clear understanding of the basic principles or real actions. The workforce and perhaps even management knew that it was only a matter of time before these corporate experts packed up their PowerPoint slides and made the run for the airport. It might almost be possible to wait these experts out.

This is not unusual; it's a pattern that has lost many manufacturers as they struggled to meet fashionable objectives rather than true improvement. On a micro scale, creating the corporate vice president of takt time is a familiar repeat of many other similar weak initiatives, all the way back to vice president of total quality, and even vice president of operational excellence.

GM is not alone in this practice. Companies seem to think that if they can name it, and slap a vice presidential label on it, they have co-opted excellence and found a way to justify or soften the blow of bad performance.

Seeking Safety in an "Exploratory" Fallback

Despite local management and shop floor support for outside help, without genuine senior management leadership, the effort was limited and powerless. Some associates landed a copy of the Lean Enterprise Institute's *Learning to See* workbook, and they were off. We'd like to report that this GM story had a happy ending, but the employees knew that in a politically charged, idea-killing atmosphere, mapping would be actually a safer fallback position than active implementation of hard improvements based on root cause analysis. So mapping it was—from one end of the operation to the other, for months and years.

Despite good intentions on the part of local management and production associates, Sharma delivered a tough message: "We would love to work with you, but sadly, we see that you have little chance for success. This process is hard, and it is better to not begin than to engage in hollow improvement activities."

Chrysler Corporation

Sometimes excellence gets co-opted and repackaged to benefit the wrong kind of leadership. One senior manager, seeing President Bob Lutz and

THE PERFECT ENGINE

Chairman Robert Eaton become enthusiastic about the LeanSigma Transformation's potential for making Chrysler very lean, decided to co-opt their excitement and score an internal political win. Using the LeanSigma Transformation as the base, the manager turned the improvement initiative into a prolonged awareness and training exercise of River Rouge proportions. First, each and every manager would receive weeks of training and awareness preparation, and then the training would cascade down to each and every employee for the next couple of years. Then the organization would be ready to tackle small sequential doses of actual improvement. Emphasis remained on getting *ready*, not getting there.

It was a safe and, from all appearances, a positive ploy intended to prolong the deadline for real action and some very hard work that would have moved equipment, changed lines, improved quality, and altered the flow of every Chrysler manufacturing facility. Putting seventy thousand employees through two years of awareness classes and training would stretch the time frame until the original objective was merely a memory, or at least until "the next thing" came along. Smart move for the manager, big mistake for the company.

Slip Sliding Away . . .

Further, the signal to the organization's early adopters, the courageous change agents who work on blind faith supported by extraordinary vision, was that while this approach was intriguing, with such a long time frame in place, it would be foolhardy to maintain a high level of interest. Best to calm down and let someone else set the pace. So the projects' original supporters, the brave and smart souls who immediately saw and understood lean manufacturing's potential, quietly shut down and slipped away. The whole movement had been co-opted.

The Operations Behind the Platform Teams

So, although the company continued to make headlines with new product platform teams and innovative cross-functional approaches, Chrysler never seriously tackled manufacturing to improve quality and productiv-

WHACK! What Doesn't Work and Why

ity. The company did not take the difficult step of making manufacturing lean. The inevitable result is that if Chrysler ever truly improves quality and productivity, the benefits will be indirect results of smaller initiatives, not of seeing the whole picture.

First fix the process, then observe and enjoy the results.

With the LeanSigma Transformation, when productivity improves on the road to lean, every quality and production problem—as well as some anticipated challenges down the road—is exposed. But becoming lean means putting in place a number of very important process trigger points—working to the takt time, minimizing time spent moving, handling, and inspecting parts, continually reducing the number of people in the process, space, and lead time.

A Fragile State of Manufacturing Process

Essentially, the good work in all these areas will make it pretty much impossible to produce or ignore bad quality; a good process simply will not produce bad parts—the system will not work. True implementation of lean principles creates a fragile state of manufacturing process that requires constant nurturing. However, because of this fragile state, all abnormalities present themselves in clear view "on a silver platter."

Leaving Management No Choice

Lean processes leave management no choice—it becomes impossible to live with bad parts quality, flawed processes, or inadequate work training. All those issues rise up to stare management in the face. It's at first glance a subtle difference between talkers and doers, the advocates of the Quality "movement" and the LeanSigma Transformation leaders who truly understand and can see where the path is taking them and follow it.

Valuing Human Assets

It is management's responsibility to preserve the most valuable assets of the corporation—its people. When managers treat people as disposable corporate assets, they are taking the easy road, rather than making the tough decisions that will sustain and transform a company over the long haul.

People are the only appreciating asset a corporation has. Everything else—machines, plants, and technology—always depreciates with time. It happens all the time, but hiring and firing and disposing at will of valuable human capital does not go unnoticed.

Leadership and Vision

Everything that is wrong with manufacturing today comes from distorted vision of senior leaders, an astigmatic distraction by little images on the periphery, leadership that has not learned how to truly *see* their operations: how material flows, how processing works, how people learn and grow, how customers use and get abused by the products.

Most leaders still look from the inside out, rather than the outside in, by keeping customers' perspective as the key driver for all their actions. Phrases like, "here is what we do," "this is what our company stands for," not "let me tell you about our customers' expectations and their problems," are dead giveaways.

Chrysler and GM would of course say that their management does not suffer from corporate astigmatism: "We're not like that. We look at people, and we understand our operations." But their actions don't support their stated understanding and vision. Organizations are complex—they do some things right and some things wrong, and many things just don't get done. Product platform teams, the reduction of product introduction cycle time, new styling, smart features, and competitive pricing are all wonderful initiatives behind Chrysler's comeback, but management failed to carry the transformation all the way through into manufacturing quality and reliability throughout the value stream.

A Second Chance for Manufacturing

The merger or acquisition by Daimler-Benz of Chrysler should, however, offer an opportunity to revolutionize the way Chrysler designs and builds

WHACK! What Doesn't Work and Why

cars. If the product platform strategy of Chrysler prevails, added to Mercedes' incredible technical expertise, the combined DNA from the two entities should produce a global technology and styling dynamo whose roots in automotive mass markets will compete head-to-head with Ford. The merger offers Chrysler manufacturing its second big opportunity for transformation, an exciting change that could affect many other worldwide operations.

One Solution for Many Problems

Fixing manufacturing and taking it forward into a leveraged position requires a sharp focus on the details, seeing the obvious possibilities, as well as waiting and watching for the unexpected ones. Amazing things happen in manufacturing; sometimes, as Samsonite associates learned, one solution solves many problems. It's a matter of observation and taking deliberate action.

In the early 1990s, one of TBM's first clients, Samsonite, was experiencing very high machine breakdown rates and had trouble meeting production quotas. This plant made card tables the old-fashioned way: Tabletops were shipped in from one building, and welding and riveting were completed at a different building. Pieces came together at yet another building for assembly and final coating and packaging.

At the end of the assembly line, four repair people kept up a fast pace, repairing welds, touching up the paint, bending a few errant legs into perfection, or tossing the unimprovable pieces—sometimes as much as 10 percent of production—into a large bin. Periodically a sweeper/material handler passed through and took away the aborted evidence of the flawed process—a strange way to make money, but operators claimed they were doing their best, and they were.

Reassembling the Scattered Puzzle Pieces

Co-author Sharma and his team decided to make their lives easier. All scattered operations were brought back into a single cell that could cut, weld, and assemble. Twelve machines and operators that were scattered throughout various buildings came together in a tiny part of one building.

THE PERFECT ENGINE

Startup in the cell produced several surprises. With all processes combined, only six of the original twelve operators could work in the cell, and for the first four days, observers noticed that the cell was functional only 60 to 70 percent of the time. When one machine broke down, the entire cell had to be shut down until the machine was fixed. The machine breakdowns were random and quite frequent. This is precisely what is intended in a lean environment. Rather than continuing to live with abnormalities, as the opportunities present themselves management must seize the moment and make the process robust.

Some members of management, however, seized on this apparent rationale for reverting to the old way: "See, it doesn't work. We've dropped from a production rate of eighteen hundred units per day down to one thousand—fourteen hundred tables at best through the cell. These cells are the real problem!"

"Maybe so," said co-author Sharma, "but the cell is telling you that you had these problems all along. In the past, when a machine broke down, you moved people around to make other parts that you didn't need and kept things going. You never really fixed the problem." Project team leaders were persuaded to bring in a maintenance person and to keep him captive in the cell. With a maintenance expert on hand throughout the shifts, there would be no delays or excuses for machine downtimes. The plan started to work.

A Shift in Responsibility

After about two days of working there, the maintenance person got restless and demanded, "I can't stand here all day. If you really want to run these machines without breakdowns, here's a list of all the things that need fixing." Management came up with the money (approximately five thousand dollars) to fix all these little problems—all at once.

From then on, the cell ran well at its eighteen-hundred-unit rate; machine breakdowns became a thing of the past. Another unexpected result of cell manufacturing was that the four end-of-line inspectors were redeployed to value-adding duties. With the cell configuration there was immediate feedback of defects and the handling damage disappeared.

WHACK! What Doesn't Work and Why

From Batch and Push to Flow and Lean

As Samsonite learned, this single transformation, from batch and push to single-piece flow and lean manufacturing, improved quality tenfold. Similarly, when the need for preventive maintenance was apparent, it brought a sense of urgency. Curiously, preventive maintenance programs have been on the radar screen for many years, one of those singular approaches that many companies continue to ignore.

Preventive maintenance, like other LeanSigma Transformation Process elements that require discipline and management involvement, is an unexciting foundation task that must be done to guarantee a perfect process. Without these unglamorous tasks, however, would-be lean producers will never reach acceptable levels of progress.

Even the adoption of pieces of the Toyota Production System—kanban and just-in-time, especially after a thorough dousing of MRP software initiatives—represented a distorted and incomplete approach to transforming an entire production system. Just-in-time became an inventory management technique, and companies rushed to adopt its apparently simple approach to replenish triggered inventory movement.

What was needed was a smooth, seamless pull system, but what many companies and gurus preached was a singularly misplaced emphasis on controlling inventory—taking it down, exposing the rocks, even in the midst of an MRP-driven push system, was purported to uncover and fix many process ills. Obviously, twenty years later, it's clear that we needed clear, unobstructed vision of the entire system, rather than narrow focus on a single element.

The misplaced emphasis was on controlling inventory, rather than realizing that what was really needed was to cleanse the system of waste and make it robust so that companies can serve customers with speed, reliability, and value that actually exceed expectations—so that companies can grow and keep their competitive advantage. That is the essence of the LeanSigma Transformation.

Lean Does Not Mean "Cheap"

The idea of making a company lean has been interpreted in its most basic fashion—lean meant "cheaper," and outsourcing production to low-cost labor markets may have resulted in layoffs at home, and downsizing or reengineered skeletal organizations, but it was difficult to argue the apparent cost savings.

The problem seemed to be, however, that all these "solutions" never improved the basic production process, at home or in China. Making and justifying these decisions, and ramping up, ramping down, and relocating entire production groups for many companies became a preoccupation rooted in break-even numbers and projections. The hard realities of how inflexible these offshore operations would be usually hit home after the fact. New product launches were complicated by logistics and cross-cultural challenges—nothing happened quickly, and there were always surprises lurching in the next teleconference update.

What the justification numbers failed to explain about true manufacturing process value is actually visible in any good lean operation, whether it is located in the Far East, South America, or New York City.

Complex Solutions, a Legacy of the Sixties

Complex solutions cannot survive; they are a legacy of the McNamarian trust in big computer solutions, an attempt to lock in all the intelligence needed to run a complex production process in the mind of an ENIAC-sized software package. It's an inappropriate fascination with software long overdue for a second look. Unplugging the systems to get the process right is the agenda for the next five to ten years of manufacturing work. Smart systems management of remote manufacturing sites will come later, but we must fix the process first.

Henry Ford knew when he set up River Rouge that good process, all the way back through his supply chain, combined with enlightened workforce policies, would guarantee him a fast, smooth, one-Model-T-per-minute pace. He knew that bad process guaranteed bad parts, bad flows, and low profits.

Ergonomics Makes Economic Sense

Bad ergonomics are another area in strong need of management attention. Our experience at a midwestern kitchen cabinet maker's plant—

WHACK! What Doesn't Work and Why

making money (and cabinetry) hand over fist, despite oceans of sawdust and unbelievably ineffective work standards—is unfortunately not uncommon. It is not unusual to see safety and health practices in many plants that require a closer look, not because they are dirty, difficult, or dangerous jobs that must be handed off to robots, but because human hands are an extension of every piece of equipment that runs a lean process. Fatigued workers or inefficient movement also contribute to bad process, and these problems may require more thought to find their natural solutions.

E-Commerce Overlaid on Bad Process

In January 2000, merchandisers all over North America sat back to count their money and lick their wounds after the first all-out dot-com Christmas. According to *Time* magazine,* for shoppers, it was fun—a new way to avoid the malls and cruise endless variety—but for manufacturers, it was a real challenge. Not every producer shipped every item on Santa's list. "If the e-commerce winners of 1998 were those with the slickest Websites, the strongest players in 1999 were those who knew how to warehouse merchandise and pick, pack, and ship customers' orders in a timely manner."

E-Supply Chains in Action

Order fulfillment became critical, and as these producers learned, you cannot win in this business without a supply chain that supports quick turnaround. Another reason e-tailers will align themselves with suppliers and manufacturers that can do quick replenishment is to make the entire supply chain responsive to actual demand and inside the delivery expectations. Not surprisingly, most had to guess well in advance how many items they needed to fill Santa's sleigh. Overestimate, and you're left holding the bag. Underestimate, and customers are putting rain checks in people's stockings.

It's a tricky balance, and it affects the bottom line. Although many producers temporarily abandoned profit objectives in favor of nailing

---

*Maryanne Murray Buechner, "How'd They Do?" *Time*, January 24, 2000.

THE PERFECT ENGINE

down market share, that short-term game plan is not one that will carry them safely through many more Christmases.

Bits and Bytes, Chips and Chunks

Amazon.com has done well to break into a market dominated by warehouse-style booksellers and publishers' extensive marketing campaigns, but behind Amazon.com's very customer-friendly façade, bad process—warehouses and forecasts and all the evils manufacturing has struggled with for years—drain profits from what is clearly a very good idea. In *Time*'s list of hits and misses, Amazon maintained the number-one position for "superior sales and customer service," along with Hallmark.com and Outpost.com (PCs and electronics); Toysrus.com, and Bluelight.com (Kmart's online entry) headed the losers list. Yet less than a week after the holiday closing, Amazon announced disappointing results, huge layoffs and massive writeoffs. Competitor Barnes & Noble.com, however, with a more efficient distribution network and logistics, remains in the black.

For surviving e-tailers, back-room processes of picking, packing, order tracking, material tracking, and perfect quality are necessary to back up Web-based order capabilities. Manufacturers who have already conquered electronic money movement and logistics tracking should have a leg up on e-tailers' processing.

Dot-com Lean

Dot-com infrastructure overlaid on inflexible or batch manufacturing processes, aided and abetted by primitive logistics networks, just won't make it, and producers know this. For years, the knowhow to set up and run lean, perfect processes, inside and outside the plant, has been available. At least twenty years of awareness and spot improvement programs packaged by hundreds of excellent training, consulting, and facilitation firms have offered producers unlimited—almost blinding—opportunities to get lean.

Further, spectacular inroads in metasystems—intelligent systems like

WHACK! What Doesn't Work and Why

the ones Yaskawa Electric uses to simulate the Bullet Train, or the system Pyramid Molding builds into its plastics equipment, or the artificial intelligence aid American Express uses to capture customer service information—are available as extensions to good, lean manufacturing processes. Yet why is it that more producers are not adopting the principles of LeanSigma Transformation?

Give Me Three Reasons Why . . . *It's Too Hard.* . . .

There are at least three distinct reasons companies don't adopt a LeanSigma Transformation.

Reason Number 1

With decades of background in mass production and absorption accounting filled with countless micro measurements of financial indices, the simple concept of producing just-in-time, in one-piece flow, driven only by customers, seems counterintuitive. Any change in operating philosophy and daily procedures, especially performance measurements—how workers keep score and how executives preserve their bragging rights—causes fear of the unknown and resistance.

Financial experts may ask, if we improve productivity by 20 percent per year, but no one is laid off, where are the savings? Even more amazingly, these same experts often say that it's great to save floor space, but reduction in square footage out there in the shop is meaningless. It doesn't reduce fixed overhead. In other words, fixed costs like capital equipment, buildings, and other more or less "permanent" elements required for production are costs that need to be distributed to various apportioned uses; having less space involved in the production process means little in that calculation.

Unfortunately, this view often wins out when a factory deploys localized kaizen improvement, without first implementing lean throughout the value chain. And if management does not proactively leverage the operational improvements into increased market share, it is impossible to provide meaningful job security.

THE PERFECT ENGINE

Reason Number 2: Short-Sightedness

Everyone loves silver bullet solutions like the "one-minute manager" and "thirty-second tire changes" and five-course meals cooked in half an hour. In fact, it is unusual to find leaders who feel that they can afford, or want, to make long-term commitments. Yet it took Toyota at least two decades to implement the Toyota Production System, and four decades later, Toyota employees are still finding that they have a long way to go.

When practitioners ask people like Art Byrne of Wiremold, Gary Christensen of Pella, Pat Lancaster of Lantech, or Keiji Mizuguchi of Showa Tekko—all of whom have been at this for many years—they will answer that they have merely scratched the surface. Even when these remarkable early adopters began to see the potential in the Toyota Production System, they were unclear about all its implications, and they took each new step not knowing exactly what was coming next. The badly translated little green book on the Toyota Production System by Dr. Shigeo Shingo was for some time their only guide, and it offered too few of the operating details—step by step instructions—that translated the philosophy into real practice.

The early adopters, however, had faith and they learned on the job. They visited dozens of plants, in North America and Japan, and they shared their findings, never expecting a nine-volume fully explicated system definition. A bit of experimentation was part of the excitement of discovering "the next new thing."

Reason Number 3: "The Fun Factor"

The third and most important reason some companies do not continue on the LeanSigma Transformation journey is that while it is easy and "fun" to make improvements, it is very hard to sustain them in a culture that continually expects the new and improved. It's a mixed blessing—once people see the power of the kaizen process, they tend to become impatient and apply this powerful lever everywhere, haphazardly and simultaneously, with unpredictable and uncoordinated results.

It's easy to see how this type of behavior, when not properly aligned to strategic objectives, can lead to the "Mile Wide, Inch Deep" syndrome.

WHACK! What Doesn't Work and Why

People see lots of isolated improvements, and they get frustrated when these improvements fail to lead to consistent, bottom-line results.

One Click Away

Supply chain pioneers like i2 and IBM are bringing the realities of the integrated extended enterprise just a click away. All that remains is for producers to perfect the processes that support e-commerce, just the way Wiremold, Lantech, Maytag, and Pella have done in all their lean initiatives.

For manufacturing, the time to make choices and fix their fragmented processes has come. Market speed and consumers demand it.

We know how to build perfect systems, how to select and train and reward powerful, lifetime employees, and we know that selective and appropriate use of technology added to perfect process will produce the winners. It's simply a matter of execution.

CHAPTER 3
LeanSigma Leadership

To lead is to mobilize and guide the energy and talent of others in the pursuit of a worthwhile end.
—*Pat Murray*

Can you imagine every product—every computer, automobile, appliance, VCR, or personal security system—ordered this morning on the Web, built this afternoon, and shipped this evening for tomorrow's perfect installation? It's a Hollywood scenario exaggerated by the popular press with little insight into actual production processes.

The production processes behind the e-commerce marketing buzz are equally important, and leaders of successful LeanSigma companies see their role as one that can, applied well, leverage the entire enterprise's success. Neglected or underachieving production processes, however, relegate the operation to second- or third-tier levels struggling for simple survival—not profits and not growth.

And yet we "know," just as we "knew" that aluminum engines are better than iron blocks, that microwave technology would quickly move from commercial to home applications, that cell phones would replace long-distance rate schemes—that product breakthroughs enabled by excellent processes have immediate and undeniable market impact. Perfect processes can transform an entire enterprise. The processes that churn out new products, as well as improved mature products, must be the LeanSigma leader's main focus.

Behind the smart new appliance and automotive offerings lies an uneven mix of production approaches, however. In the worst plants, desperately lurching systems push an erratic product flow out. Producers in the Islands of Excellence enterprise seem to effortlessly move from one model to another and from one new feature to the next best edition. But in between, in that limbo region where squeezed producers struggle to compete, the everyday frenzy of "getting there" churns up the marketplace and regularly burns through the contenders who did not quite have it—the right process, the perfect delivery system, or the lines of new launches ready to release.

The Fallacies of E-Business

Unfortunately many companies are learning the fallacies of Web commerce: Setting up a Website, taking orders, and shipping overnight do not guarantee profitability. Christmas holiday sales in 1999, the first big e-tail season, reached $3.5 billion, according to the Gartner Group 2/2/00 webletter. And it was not until the holiday registers had shut down and the returns were back in that eager Web innovators began to see the red ink and backorder problems, but many of them were unclear on the real causes. What happened? Were the manufacturers unprepared? Was the Website crushed? Or did we simply miss the hot toy every child demanded this Christmas?

The answers are actually rather simple, but the solutions for some companies will come too late. For many companies, Web commerce transactions are unprofitable because of back-end fulfillment problems, some as far back as basic production performance. The Gartner Group recommends that "rather than rush headlong into Web commerce, companies that want to sell physical goods over the internet should first develop a strategy for efficient back-end fulfillment."* Easily said! "Enterprises that fail to do so will have lower profitability and may permanently damage customer relationships."

Enslow cites seven major fallacies, tripping points, that even the best e-tailers must consider before they go "live."

*B. Enslow, "The Fallacies of Web Commerce Fulfillment," *Commentary*, May 1, 1999, Gardnerweb.com.

THE PERFECT ENGINE

Fallacy #1: We're selling over the Web, so we must be making money over the Web.

Response: Revenues do not equal profit, no matter how many hits are on the server. Gartner points to the unpleasant fact that when Web commerce transactions are unprofitable because of back-end fulfillment problems, and when enterprises fail to fix the back-end, they risk lower profits, but more damaging, they risk destruction of customer relationships, the company's future.

Fallacy #2: Our differentiation on the Web will come from our front-end processes.

Response: Think about other differentiators, because when everyone is on the Web, your company needs to look different as it delivers more. The addition of experienced personal, technical, or customer engineering specialists, something Dell is expert at on their Website, in conjunction with Web info is most attractive to many customers' decision process.

E-commerce leaders now want to differentiate themselves through better flows, and faster and cheaper deliveries. Again, companies that want to sell like Dell over the Web must make fundamental changes to their procurement, *manufacturing,* and distribution processes.

Fallacy #3: We can always do back-end integration later.

Response: Wow! The eternal optimist reigns! Exactly how much later? And will customers wait, and do they care? Gartner's newsletter recounts rumors of legions of temporary workers hired to rekey data from a Web ordering system into a traditional order management system. Customers on permanent hold, unsolicited orders, duplicates, lost backlog, out-of-stock offerings, and other horror stories take on mythological power. "Don't go there, they can't deliver," and, "If you put me on hold one more time I will disconnect," are not the kind of customer feedback comments your company wants to receive in its e-mail bag.

Fallacy #4: Our logistics operations can handle Web commerce fulfillment.

Response: Prove it!

Big brick warehouses and dozens of fork trucks and shipping departments geared for high-volume full loads were not designed to ship thousands of overnight small bundles. Over Christmas 1999, a Massachusetts high-end clothing retailer discovered how much stress wide-open e-orders could put on a system staffed for big shipments to stores, rather than smaller ones to live customers. Despite the addition of sixty temporary workers, all of whom had some training, each day of the last two weeks before Christmas was retailer hell—unfilled orders, bad dates, and mixed-up trucking schedules. They barely survived what should have been the best time of the year.

Fallacy #5: Our existing supplier relations will support Web commerce.

Response: Oh really?

Adversarial relationships and blinding focus on cost reduction won't get retailers and their supply base through the systems transition to e-based replenishment. Innovative systems and a shared desire to "be the first" will, however. In some cases, as with EFTC, retailers may want to arrange direct shipments from suppliers to end customers, taking the e-tailer out of the loop, except for electronic receivables collections and design.

Fallacy #6: Our order management system can handle Web commerce.

Response: Un-huh . . .

Order management systems, like warehousing and shipping systems, tend to be designed with certain activity levels in mind. There is no quicker way to incinerate your glorious Web business strategy than failing to provide adequate front-end order-processing power in advance of pulling the switch. Dozens of companies also learned this lesson over the holidays, and they have not recovered.

Fallacy #7: We can now sell effectively to anyone around the globe.

Response: Global exporting is a bigger and more expensive exercise than setting up shop at the mall. Lead-times, customs and freight forwarding procedures, and international carriers represent a new learning curve with higher risk.

Experienced international trade experts may be your best planning tool.

For managers, it's a traumatic, energized, and out-of-control battleground; for production workers, it is a life-shortening grind. But for the visionary leaders, the winners who manage to outgrow and shed heavy mass production structures, it's an exhilarating ride, an exercise in perfecting pure process, an experience that elicits exhilarating cries of postgame celebration.

THE PERFECT ENGINE

Learning from the Masters

Leadership in the new economy with its constantly changing technology landscape requires an ability to adapt the new and emerging business model while maintaining constancy of purpose and fundamental values. This ability is enhanced if the leaders:

- Make deliberate choices that are consistent with the values and beliefs of the enterprise.
- Learn to protect the voices from *below*, the ones without authority.
- Have followers who have the choice *not* to be.
- Have a unique ability to distinguish between what is precious and what is expendable to stay relevant.

Today's business leaders must have fundamentally sound strategic principles and core values, but they must adapt the business model according to the changes needed to satisfy emerging customer needs. Digital evolution provides quicker, direct feedback, but not necessarily intelligent insights. Listening to the customers is, therefore, even more important for insights and new directions. In fact, leaders must seek out and listen to some of the most demanding and challenging customers. In an increasingly customer- and consumer-centric marketplace, customer experiences and transformations are new economic currency in addition to the goods and services. Successful leaders must constantly reinvent their business model to provide positive experiences and lasting transformations.

Certainly, global worshipers of lean systems have the benefit of over eighty years' hard work by Henry Ford, and the post–World War II Japanese recovery teams dominated by Taiichi Ohno and his disciples, to draw on.

Lean kaizen pioneers like Lantech, Critikon, Pella, Alexander Doll Company, and Mercedes each developed a particular system of custom kaizen approaches as they moved their processes from the ranks of early adopters to seasoned lean practitioners. Their innovation and hard work have taken the lean methodologies beyond Toyota Production Systems' initial efforts into techniques that work in small and large companies, with nonunion as well as union workers, and in Western environments that tend to be more worker-enabled than those of pioneering Japanese methods.

Leadership Makes All the Difference

The pioneers did not, however, achieve exponential successes in relatively short time periods armed simply with a book and a pep talk. Lean leaders, professionals like Lantech's Pat Lancaster (one of *Industry Week*'s "Kings of Kaizen"),* Art Byrne of Wiremold, Gary Christensen of Pella, Karsten Weingarten of Mercedes, Herb Brown of Alexander Doll Company, Carole Uhrich, formerly of Maytag, and others, have wrestled the beast for almost fifteen years, and despite results in the double- and triple-digit range, they're not done yet.

It's not enough, however, to learn from the pioneers and to expect that pure process simplification and perfect quality will carry transitional producers into twenty-first-century speed and responsiveness. New technology leadership challenges are all around because the merger of Web communications with well-managed and consistent production systems will remain an industry opportunity for the next ten to twenty years.

Building and Leading the Innovation Machine

What is the magic that enabled the pioneers to recognize a powerful solution when only fragments of a pixelated image were presented to them years ago? And what can today's executive learn from their breakthrough initiatives that will carry the millennium organization over the successes and inevitable mistakes into the next generation of technologically enabled lean process?

Take out scrap, rework, rejects, and waste, and what remains is process integrity.

Lantech

Pat Lancaster, now chairman of the company he founded in 1972, believes he has developed some insight into the transformation process: "It is like taking a Band-Aid off your hairy arm. If you do it fast, it might not

---

*John H. Sheridan, *Industry Week,* March 2000.

THE PERFECT ENGINE

be perfect, but now you're talking about how bad it *was*—not how bad it is *going to be*."

And for this lean pioneer, the results keep pouring in. Over an eight-year period, since the company's first kaizen steps with help from TBM, sales have more than doubled, while the workforce has remained steady at 320 employees. Company-wide inventory turns jumped from three to fifteen and rising. The manufacturing cycle time (throughput time) is down to *eleven hours* from twelve weeks or more pre-lean.

And in one of the most impressive lean improvements cited by most of the early adopters, Lantech has come to enjoy a company-wide productivity gain of over 17 percent per year for eight years, for a cumulative productivity increase of over 100 percent.

All of which has fueled what distinguishes Lantech's remarkable technology transformation—from a position of near paralysis, floundering after key patents expired, to an incredible new product introduction rate. Lancaster's battle to create newer and better customer solutions has produced an innovation machine, a gem that is pointing the way for many slower, technologically challenged giants.

Pella

Mel Haught, Pella's senior vice president of operations, executed the hard work of leaning down factories and lines to facilitate exponential variety and brand differentiation in the market. It is hard but necessary work that requires Lean Transformation missionary leadership.

Pella, the Iowa producer of wood windows, enjoyed a pre-lean situation that was healthy, but without significant growth. Pella was a smaller player in a bigger market. Profits were good, but Haught was not satisfied just with good profits. He was drawn to the leadership challenge of making this solid family-owned business a better company, and he felt that there was so much opportunity there, as well as native discipline and technical skills, that he knew it was time to take it one level up.

Haught evaluated the flow of value through the entire value chain, looking at suppliers and distributors for possible opportunities to shrink and compress throughput times. In fact, he has issued a challenge to deliver an entire house (of windows) at one time within one week of order-

LeanSigma Leadership

ing—an unheard-of but technically achievable challenge that would put contractors and homeowners in construction heaven.

Maytag

Maytag leadership is equally excited by the new market opportunities that their innovation machine has generated. Carole Uhrich, former executive vice president, member of the Maytag board of directors, and president of the new Home Solutions Group, believes in the power of transformation to offer leadership fresh opportunities for market dominance and profitability growth.

Uhrich describes the impact Maytag's new product strategy has had on the appliance industry. "Maytag has created a unique place for itself in the marketplace," says Uhrich. "It was not known for innovation. Maytag has changed that. We are delivering innovation to the upper end and to popular price points."

The Ultra-Premium Strategy

Maytag has raised the bar on the value and premium product classification scheme by adding a new group to the top, the ultra-premium brands, starting with the 1997 introduction of the Neptune washer. Priced at more than $1,500, this model eliminated concerns that consumers would not pay this high price for a washer. And the ultra-premium idea apparently took—production lines at Newton, Iowa, have been at capacity for this model for over three years since its launch. Consumers readily accepted it. Despite the higher up-front cost, consumers apparently felt that the innovation value offered behind the premium brand and the annual utility cost savings more than compensated for the additional cost over the life of the product.

A second ultra-premium product offering, the freestanding twin-oven Gemini range, again confirmed the marketability of ultra-premium products. Priced at $1,399, with two ovens in the space of one, this model is selling very well to consumers who will pay for flexibility, extra features, and extra capacity.

In February 2000, Maytag introduced a third market-making product

THE PERFECT ENGINE

in the ultra-premium category, the Maytag Climate Zone refrigerator, a product that is featured to optimize the consumer's food storage and preparation options. For about $150 more than the average high-end unit, consumers can custom control different refrigeration zones for their meat, produce, and milk. Climate Zone was designed to offer flexibility that improves taste and freshness of foods while it saves consumers money.

Powering the Innovation Machine

Strategically, these three new designs are just the type of innovative product that can change a product category and, at times, even an industry. But creating waves of innovation and filling the new high-end market niche is not without its production, procurement, and logistics challenges, just as it would be for any new startup.

Following Maytag's high-growth strategies for new markets with operating excellence is critical. "It is," says Uhrich, "not sufficient to just have a source of revenue growth. We have to deliver a wide range of products, and we have to be positioned more competitively from a cost point of view. That leads us to LeanSigma, Maytag's total manufacturing excellence program that combined the best of lean manufacturing principles to eliminate waste and the best of Six Sigma principles to deliver quality. All design engineers are being trained in LeanSigma principles, so those future product introductions will be consistent with operating excellence standards.

Corporation-Wide Changes to Support Market Needs

E-commerce will change the way businesses take orders and contact customers, but not all Maytag products are immediately affected by the e-linkage initiatives. Says Uhrich, "As you go forward, you can clearly envision a point in which all transactions are paperless, without a person putting anything on a piece of paper. Orders and order fulfillment, for example, will be Web-enabled, and we need different software and different manufacturing configurations as well. That's why we have the

LeanSigma Leadership

make-to-order and make-to-replenishment initiatives in targeted areas of our businesses."

"But," warns Uhrich, "this is not the kind of thing to implement massively across the company—lots of products will not be made to order, a lot of what Maytag sells through Wal-Mart, Sears, or Home Depot, for instance, would not be made to order. But we need the flexibility to operate in a number of markets."

Pushback

Not all good decisions are equally well received within an organization. Some amount of pushback is to be expected, "And in fact," says Uhrich, "some problems are legitimate. You have to help people get through it. People have a lot of different things on their plates." In the engineering area, for example, "We can list a hundred things that need to be done, and if someone says, 'Look, I can't get all this done,' sometimes we have to prioritize, or bring in more resources. Engineers are key to this transformation."

It's a big change in an industry that hasn't moved as quickly as others, such as consumer electronics and computers. The time clocks in Maytag's transforming operations run faster. It's a significant change for a primarily domestic industry that never had this level of international competition.

Leadership for LTP

LeanSigma Transformation leaders have demonstrated excellence in all sectors of the business, but typically they first started to work on one key system segment and spread the initial successes to other areas once the fire started. It is important to pick the right spot and to deploy resources deliberately where management is certain they will have maximum impact. Leaders understand that to avoid dissipation of valuable energies, teams need to be focused on a few critical efforts, rather than deploying in dozens of directions all at once. *Deselection becomes even more important than the selection of the vital few to focus the enterprise energies in the right direction.*

THE PERFECT ENGINE

There are three opportunity areas for lean leadership:

1. Leadership for internal processes and people
2. Leadership for the supply chain
3. Leadership for market responsiveness

1. Leadership for Internal Process and People

Certainly such companies as Toyota and Mercedes became leaders in pure perfection of production processes long before they shifted these powerful energies over to suppliers and the marketplace. Once associates have learned to observe and make changes in assembly of key components, for example, they will find that the same techniques work in engineering and other white-collar areas. Shop floor is the ideal place for training and visualizing most of the lean principles for the biggest, sustained impact.

EARLY ADOPTERS

Early adopters had vision and trust, and they were well-recognized and frequent communicators of the simply amazing power of the new kaizen methods. They were charismatic and legendary figures whose reputation preceded them, but they had constant struggles with culture change. When the early adopters of kaizen methodologies first concentrated on opportunities on their own shop floor, they wanted to improve quality, predictability, productivity, and speed in basic processes.

Leadership in early applications focused on getting the methodology right and in taking the enterprise deeper into a more powerful, long-term process change.

First projects typically included creation of one-piece flow, setup time reduction, housekeeping and reorganization of tools, and improving parts presentation areas. Setup time reduction projects became attention-getters that opened people's eyes to the exponential improvement opportunities. Teams at Wiremold in Hartford, Connecticut, for example, proudly pointed to huge time reductions as they strengthened their collective skills and their new team-based work groups.

LeanSigma Leadership

Soon, team members and executives learned from doing and observing flawed processes that the possibilities were endless.

Leading Cultural Transformation

Perfecting internal processes is time-consuming but, given the will and determination, it can be accomplished. The hard part is to change the organizational culture and the mindset of the people, so that the new way becomes their way of life.

This cultural transformation requires leadership and constant nurturing to keep the weeds of doubt from taking over the new lean garden. It requires the ability of the leadership to develop a clear, concise, and compelling vision of the future state and then communicate that vision with passion, enthusiasm, and repetition until it becomes part of the DNA of the enterprise.

If you are comfortable, you lose your alertness.

Teruyuki Maruo

Leading by Example

Change leadership requires leading with examples and gut level conviction, rather than simple words. This is why part of the lean leadership development process is hands-on experience of the weeklong process by the senior leadership.

Bill Kassling, CEO of Wabtec, after his first kaizen experience in 1991 decided that this was the best way for him to learn about his people and processes and to demonstrate commitment. Over the next five years he set aside ten weeks each year to participate in a weeklong kaizen event in one of his company's plants.

Art Byrne, CEO of Wiremold, used meetings, poetry, posters, cartoons, and other attention-getting devices to signal his unyielding intention to proceed with big process changes. One inventory reduction exercise was preceded by his announcement that by a specific date, he expected certain fenced-in parts storeroom areas to be eliminated; Byrne followed up the announcement with a sign affixed to the newly padlocked storeroom fixing the imminent date of final inventory disposition. His first exchanges with stock-keepers and production workers rein-

THE PERFECT ENGINE

forced this preliminary attack on inventory, and soon the word was out that Byrne meant business. He was unwilling to delay or retreat from Wiremold's lean momentum, because very early on he recognized the benefits of momentum in transformation.

MODEL BEHAVIOR

Leadership in the eyes of employees is critical at very early stages of any new transformation. Employees are quick to notice management's lack of support or even lack of understanding; for their own protection, employees quickly reach their own conclusions about the viability of the new approach, and their willingness to join in is directly proportional to the strength of leadership conviction evidenced by executive management. Anything less than open and visible support will shortchange the process and can lead to employee disillusionment.

Essentially, strong leaders learn early on to model the behavior they seek—skilled listening, communication meetings, data sharing, teamwork, and rolling up their sleeves to lead by example—to get the characteristics they want in the new organization.

BAD MANAGEMENT DOES NOT GO UNNOTICED

Likewise, management's failure to visibly support and model new behavior is the most frequent killer of good kaizen culture changes. In one high-volume office furniture producer located in the southeastern United States, kaizen events were scheduled for every week; launch activities continued to build excitement, as employees anticipated new ideas that they believed would mean radical improvements in the way they worked. Team leaders attended prerequisite lean training sessions as supervisors hopped in their cars to tour advanced lean sites. The kaizen newspaper and kaizen war room soon filled with project ideas and schedules. A startup project only served to whet shop floor associates' appetites as they cleared the decks for twelve-hour workdays ending in pizza suppers.

But management, believing they could delegate this kind of intense process and culture change, hung back and waited for results. Although the plant manager was seen in various parts of the plants, silently traversing final assembly and feeder areas, production workers heard little and saw less from "supportive" executives. Little by little they got the message: It was a wait and see exercise.

Not having truly demonstrated or committed their energies to lean initiatives, executive management had decided to take the position of least risk—hanging back and watching. They became observers, and potential finger-pointers, rather than leaders who modeled new behavior by their own example, or participants in early morning team organization meetings. They chose not to expose themselves.

In a sense, management unknowingly modeled the kind of behavior that weakened and stalled out employees' enthusiastic first efforts. "Wait and see" turned into "don't try too hard," "don't stand out," and "don't rock the boat" as shop floor passions faded—a somewhat predictable outcome.

2. Leadership for the Supply Chain

Suppliers are in the habit of mistrusting big customer efforts to force quality, cost, and delivery goals upon a captive supply base. They are inherently suspicious of leaders who may speak in terms of partnering and trust, but behave differently by demanding precipitous cost reductions or impossible deadlines for new product ramp-ups.

The Lopez Method—a heavy-handed, strong-arm approach—did not work for GM in the long term, although short-term it garnered the auto giant millions of dollars in savings. The problem with this approach to improvement is that coercion, dishonesty, and heavy-handed leveraging tend to put smaller partners, possible sources of innovation and improvement so clearly needed by larger, slower members of the enterprise, out of business—or at minimum, out of reach.

Leaders may have framed their new relationships in the language of partnership and trust, but suppliers, long accustomed to heavy-handed approaches framed in corporate double-talk, know a bad leader when they smell one.

SUPPLY CHAIN MUST BE SEAMLESS

Leadership in the supply base, therefore, must be based on communicating consistent messages, information sharing, and sometimes sharing of train-

THE PERFECT ENGINE

ing and kaizen breakthrough improvement resources. These are the elements that prove the strength of the vision. It's all part of truly participating in enterprise leadership initiatives, rather than taking the Lopez coercion approach. Pella, Mercedes, and Chrysler have been especially successful at tailoring and encouraging lean initiatives among their suppliers.

Supplier integration in the extended enterprise runs on a continuum from awkward efforts at simply building communications between customers and suppliers to sophisticated technology and savings sharing worked out between production leaders.

LeanSigma Transformation success stories, however, shine among assemblers as well as suppliers. Pella, for example, under the leadership of Mike Buchheit, has successfully exported kaizen techniques to a number of critical suppliers—Cardinal Glass being one of the first.

SUPPLY CHAIN BENCHMARKS

All producers wish they were blessed with an enterprise that maintains consistent quality and delivery performance levels at all tiers of the supply base. What they fail to recognize, however, is that excellence comes in many forms. Small and medium-sized suppliers' journey to lean high performance levels will not be the mirror image of the large customer's operations. Smaller suppliers tend to be more adept at adopting appropriate new technology; they are by nature lighter and more agile, a smaller and somewhat younger mass to turn around and move in new directions.

There are, however, certain basic benchmarks common to both customers and suppliers that identify strong performance, and these benchmarks above all should be the drivers of change, as well as the relationship builders. They include:

1. On-time delivery: Material received on time, not early and not late, is considered as a quality receipt. Some companies set guidelines to allow for early receipt, but no high-performance enterprise can tolerate late deliveries. Both early and late deliveries are indicators of bad process.
2. Quality: Quality as measured by field performance or performance at final assembly is a preferred measure compared to detailed

component and subassembly quality measures. However, certification of suppliers to indicate that their process and products consistently meet specific quality levels has, especially in the automotive industry, reduced the number of tests and the amount of data gathering required to reinforce excellent quality performance.
3. Responsiveness: In the new economy responsiveness is as important as quality and delivery. Suppliers must become an extension of their customers' capability. As the demand for customization and new product development increases for all manufacturers, suppliers have to become agile and responsive to new product introduction teams and make contributions early on in the process.

Supply Chain Responsiveness

The electronics industry suppliers are perhaps the best source for comparative evaluation of seamless cooperation and high-performance execution. At Flextronics International, for example, the Santa Clara producers of electronic equipment for customers such as Cisco, Bay Networks, Hewlett-Packard, Apple, and others, associates have honed their new product ramp-up and procurement skills to a point that anything slower is boring. Flextronics and other electronic outsourcing experts like them—Solectron, Celestica, and SRI—are essentially in the new product introduction business, and their processes and leadership are tuned to a different rhythm.

Michael Marks, Flextronics' visionary and peripatetic CEO, epitomizes this multitasking approach that pushes internal capabilities to new heights every day. And one of the very noticeable keys to Flextronics' rapid growth to number four in its market is Marks's ability to leave the organization loose and flexible to make rapid changes, and to bring acquisitions and new business initiatives into the main organization without classic new versus old rivalries. Marks signals through his strategic growth efforts that the best way for Flextronics employees to perform their work is without boundaries or rigid task descriptions. In fact, he models that ideal growth-enabling behavior by being a leader for organic change. Further, as a chief executive who recognizes the strategic leverage obtained through high-performance manufacturing, he continues to make production a core competence.

THE PERFECT ENGINE

3. Leadership for Market Responsiveness

We've mentioned Maytag's transformation in several spots in this book—its strategy, its process redesign, its incredible use of simulation to design new processes, and its approach to building an innovation machine. We believe that unique leaders have multitasking capabilities—they work with their eyes on the horizon, even as they push and pull pieces of the organization along a new path. Lloyd Ward, former CEO of Maytag, is just such a leader, one who simplifies issues in order to communicate his vision through simple images.

Ward is a clear thinker who processes information at high speed, all the while ingesting an enormous volume of input data, which he parallel processes to generate a clear picture. The next step is communication to all levels of Maytag and the outside industry and financial watchers.

Ward's special talent is an ability to take the future, which looks very cloudy, to remove the smoke so that we see clearly, and to wrap the message in a credible corporate voice. And as one close associate sums up the impact of his leadership style, "We will follow him absolutely anywhere, anywhere."

One of Ward's key rallying messages is that the appliance industry must find another way to grow revenues—the dream of unlimited growth is gone, a nineteenth-century expansionist relic whose natural successor is product growth and market segmentation. The new growth model puts process demands where production leaders have felt no special pressures in over twenty to thirty years.

In the appliance industry Ward translates this market shift away from simple consumer purchase and replacement decisions made when appliances break down. In essence, dependability, Maytag's corporate logo, grew the company's initial market position, but stalled in the twenty-first century. Paradoxically, as appliances have become more dependable with fewer breakdowns and less wear, consumers can delay purchase decisions beyond obsolescence issues.

CREATING A MARKET FOR "WANT"

Ward envisions technology-rich product offerings creating a new growth built on features: "We want to fuel a new growth strategy that responds to consumers' wants and even unstated desires, rather than simple needs. New features in new appliances create new growth opportunities."

The Gemini two-oven range, for example, illustrates family and meal preparation changes that are best served by multifunction cook units. Another brainstorm: Smaller, two-person families won't fill a dishwasher even with a whole day's worth of dishes. Why not change the appliance design to fit changing consumer needs? It's a novel approach, but one that is true to lean principles, follows the consumer, and responds—a perfect, silent, unbidden servant to the customer's psyche.

Ward is a leader who wants to understand, anticipate, and move with the kind of yet-undiscovered changes that will move his industry closer to consumers. The average harried food preparation process, for example, just to assemble a passable spaghetti dinner, takes over four hundred steps from start to finish, a kaizen challenge calling out for simplification, and the first step is designing and positioning the tools (appliances) in the work cell (kitchen) to smooth product flow (food preparation).

Process mapping the kitchen work area will reduce steps from four hundred to one hundred; redesigning the positioning of utensils and appliances will further improve the ergonomics of the meal preparation and cleanup. By thinking through the customer's daily tasks, Ward understands that new LeanSigma opportunities will appear for Maytag, market-making opportunities to let technology solve everyday problems.

Ward's approach to innovation leadership means taking a small piece of valuable demographic information—family size, for example—and building an entire world from it. Designing different doors to accommodate smaller dishwashing families, for example, or offering sanitizing as a finishing step for moms who pop pacifiers into silverware racks, illustrates how demographic and valuable focus group feedback can affect an entire design concept and take a market down a new path.

LOOKING INTO THE FUTURE, THE SOLUTIONS-BASED ECONOMY

The nuggets that Ward's innovators mine to create valuable new appliance features don't come out of the sky, however. Ward hires experts to study the landscape, to follow homeowners through their day, observing

THE PERFECT ENGINE

their movements and retracing steps with the sharpened eye of a kaizen master. Marketers call it deep research, but Maytag calls it understanding their customers, going to the heart of their lives.

Ward's communications strategy is based on sharing the vital corporate statistics—financial indicators like price to earnings ratio—with the market, with suppliers, as well as with employees who make it happen. Periodic online discussions with market analysts are broadcast simultaneously to all employees, a leadership method that builds internal focus on real profitability and growth challenges.

Maytag's competitive response to cost-based market strategies is to introduce new products with remarkable features before the market recognizes need. This type of market leadership of course calls for a flexible and fast underlying production and new product engineering system. The LeanSigma projects that Maytag has undertaken in production and in the Design for LeanSigma lab continue to enable the company to morph from traditional, dependable appliance manufacturing to startlingly innovative product introductions like the Neptune washer, for which Maytag received thousands of orders on the Web, long before the product had even been released, and about a year before Web-based ordering was commonplace.

LANTECH

Lantech's CEO, Pat Lancaster, has taken the same innovative approach to planning for his company's future. By making dozens of visits to customer sites, like Kellogg, Lancaster can see and study in person exactly what end product his customers are working to make faster and better. That kind of down and dirty field research is the only way to truly understand big customer processes and to incorporate innovation in the next shipments.

PELLA

Pella's president, Gary Christensen, has introduced innovation on the marketing side at Pella, an area that carried a certain aura of staidness and inflexibility. When Christensen decided to create brand differentiation in the company, he started with an architectural series, a designer series, and Pro-line, and by branding these window products differently, he created differentiation in the marketplace based on special features. The new marketing strategy has created hyper growth in a slow-growth industry. This market leadership has taken Pella to a new level of revenue growth of 20 percent per year in an industry that is only growing 4 to 5 percent per year. As a result Pella has become one of the two biggest players in the high-end window market.

At Maytag and at Pella, taking a nearly one-hundred-year-old institution and injecting the voice of the customer around product use and unfulfilled wants is at the very creative heart of the new innovation strategy.

How to Build Leadership Strengths

Each of these three leadership challenges—internal process, supply base, and market responsiveness—demands that leaders rethink traditional leadership practices. Leaders for the LeanSigma Transformation have come to understand that overall, their concerns must be with perfecting all the enterprise processes, from production processes that affect throughput and cycle time, down through supplier links that pull perfect product through Web-enabled links directly to customers' needs and wants.

But carried beyond internal corporate goals, leaders like Ward, Lancaster, Christensen, Haught, and Byrne are driving innovation throughout their enterprises, creating new industry standards for breakthrough thinking.

Manufacturing Leaders Climb into the Driver's Seat

For decades manufacturing leaders have complained about being second-class corporate citizens, with less visibility and less perceived power inside their own corporations. All that changed in the 1990s as companies discovered that perfect processes that result in improved quality, cost, and delivery—all the domain of operations—could make the difference between profitable growth and mere survival.

THE PERFECT ENGINE

It is important to understand that each of the leaders profiled in this chapter took a dominant process-based position early on and carried out strategies dependent on excellence in their production processes. These executives every day live out the belief that absolute responsibility for leveraging their corporation's success rests in their hands, and it is a gift, an opportunity they must exercise daily.

Lessons for Future Manufacturing Leaders

For any manufacturing professional preparing for the same earth-shaking shift, here are four guidelines that will increase your value (and visibility) to top management:

1. BECOME A PROVIDER OF SOLUTIONS

Executives love managers who arrive with solutions, not more problems. Statements like, "We've got a problem out on the press and we ought to have it fixed by this Friday," or, "We are ready to go to launch, but it's gonna cost you," are bad messages that put the bearer in the negative position of bringing *one more problem*. Instead, become a provider of *solutions*.

Couching your messages in the language of solutions—we had a problem, we took action, and the solution is ready—is what you must practice.

2. PURSUE PROCESS PERFECTION RELENTLESSLY

Leaders understand and expect the kinds of shifts and reforming of organizations that morph around process. Everything in LeanSigma and kaizen is designed to uncover and cleanse true process, and to protect this profit-making tool for creating and launching new products, new features, and new ideas.

3. COMMUNICATE YOUR COMPETITIVE PERFORMANCE IN NUMBERS

Although internal measures like throughput time and inventory reduction results are impressive, what the market and future customers truly care about is quality, price, new features, and time-to-market.

All the other numbers support process improvements that will take your organization far on its journey, but leaders understand that different listeners—financial analysts, consumers, suppliers, technology partners—listen to different numbers.

4. Communicate Success Indicators Enthusiastically

Learn to selectively broadcast the right performance indicators as you work at building the strength of the enterprise in all critical areas. With greater visibility come resources—better new hires, bigger budgets, more training, higher salary packages, and better tools. By truly understanding and communicating manufacturing process's contributions to the enterprise, new leaders put themselves at the head of the race for position, resources, and power against marketing, engineering, and supply chain competitors.

Internal change for the workforce, as well as the production process, requires new leaders, and the corporate pioneers whom we have introduced, from the beginnings of the lean and kaizen movement down through the e-economy, maintained their change focus through many challenges.

Early adopters, innovators, and supreme implementers alike share one common characteristic—they all, from the beginning, understood the importance of their very visible, participatory role in creating success and leading their organizations through many difficult changes. As enlightened leaders they understood early on and believed completely in the power of manufacturing to create success and power in any corporation, and with great determination they courageously drew on the right ideas and human resources to execute the dream.

Leadership for the Experience Economy

Leadership for the experience economy requires cultural transformation to lead the organization to growth in the new millennium. Leaders in tomorrow's manufacturing world must be flexible to survive in constantly changing technology and business landscapes.

They must also, at the same time, be firm and consistent in enforcing adherence to the fundamental strategic principles and value systems that keep the enterprise energized and that add value to society.

THE PERFECT ENGINE

> **Emerging Leadership Model**
>
> To deal with adaptive challenges of today the old model of productive, obsessive experts is ineffective. New economy leadership requires:
> - Ability to protect voices coming from below without authority
> - Ability to have followers who have the choice not to be
> - Ability to distinguish between what is precious and what is expendable
>
> **"Who vs. What"**
>
> ---
>
> **Emerging Business Model**
>
> You must have fundamentally sound strategic principles and core values but the business model should be adaptive.
> - Digital evolution provides quicker, direct feedback but not necessarily intelligent insights
> - Listening to a customer is even more important for insights and new directions
> - Seek out and listen also to three of your demanding and challenging customers
>
> **"Reinvent Your Business Model"**
>
> ---
>
> **Emerging Marketing Model**
> - Customer experiences and transformations are new economic offerings in addition to commodities, goods, and services
> - Make discoveries by listening to your customers—what keeps them awake at night?
> - In the new information-sharing economy brand equity becomes even more important—you must walk the talk
>
> **"Your Customer Is Your Product"**

Figure 3-1 Emerging leadership model.

These kaizen leaders—people like Ward, Byrne, Haught, Christensen, and Lancaster—became legends because each one of them has an unusual and powerful approach to change. In any era they would have been change agents, and in any industry they would make a difference, but in manufacturing today, they are continuing to lead a revolution into our next big transformation, the solutions economy.

CHAPTER 4
Preparation for Transformation and Innovation

Quick and Simple Beats Slow and Elegant

Why preparation is so important:

Sometimes companies seize an opportunity to change the way they work because of market pressures; sometimes companies, having simply reached another growth stage, need help to reach the next rung on the ladder; and sometimes very specific internal issues drive the organization to understand and reluctantly accept the need for massive change. Xerox, for example, is facing all three challenges—market pressures, the need to reinvent its technology base, and internal profitability issues. Lantech and Wiremold reached similar turning points, and their dedicated and visionary adoption of kaizen and the LeanSigma Transformation took them to new levels of success, profitability, and growth. Pella, on the other hand, was profitable and did not have to change. Pella management, however, wanted growth and market leadership.

But the kind of transformation and organizational change that is achieved by dedicated LeanSigma practitioners requires a paradoxical combination of energy and excitement, as well as pain. And how should companies in what we call "the second wave," adopters who will find some benchmarked examples and guides ahead of them, approach the formidable possibilities offered by LeanSigma Transformations?

The answer is twofold—companies will do better, and generate more

THE PERFECT ENGINE

lasting results, with the right preparation, and with an awareness of what's ahead. Although we cannot predict every single step and each reaction that your company will experience on its journey, we know from twenty years of building this methodology that organizations can expect excitement, confusion, some pain, and huge growth spurts as they progress.

Organizations and individuals as well must prepare for change and welcome the disruption and questions that transformation inevitably brings. Awareness education, technical training, hands-on experience, strong and visible leadership, unleashed creativity, and a few other proven techniques will make the transition easier.

Change Can Be a Positive Experience

Change—any tectonic shift in the company's landscape—always brings with it new opportunities. Some of us think that movement, any kind, is good. IBM, for example, facing the end of its hardware era, was able to turn to its long-held internal creativity to become a solutions provider that captured market share in a wider marketplace. Now, after nearly twenty years of rebuilding, IBM again is taking a dominant role in hardware as well as software solutions: The company has come nearly full circle. IBM's transformation was necessary, and quite difficult. Lantech's unfortunate loss of certain patent rights caused the company to focus on its product innovation skills, a competitive strength that has taken it again to the height of its market sector.

Change spells opportunity, but every lean practitioner must be prepared to reinforce excellent basic processes with new waves of creativity. Bright ideas don't last long in the open when they are not grounded in perfect process, well-oiled operations that can deliver.

Generation-Long Change

So the challenge for organizations new to lean and kaizen, and for all advanced practitioners, is to preserve sound process, as kaizen continues to release waves of creativity and innovation to completely transform the

Preparation for Transformation and Innovation

way work is done. Truly, this is not a casual or short-term commitment; it is a lifetime change not unlike creating and educating a whole new generation of leaders. Although leaders might expect, even demand, to see quick results, for long-term realization of full benefits, plan to continue to invest in your company's transformation. It never stops. It's a journey and not a destination.

Managing the Change to a Lean Culture

The change from traditional production methods to lean processes should, to be truly successful, transform every element of the operation—from the way plants are laid out to the way results are measured and the way associates are trained and rewarded. Specifically, the following elements of the organization will have to be transformed to support lean operations:

- Information systems
- Supply chain relationships
- Cost accounting system
- Incentive systems and management policies
- Organizational structure
- Workforce involvement

Information Systems

Lean practitioners report that the biggest change they encounter all through their operations is the shift from "push" to "pull" methods. Pulling product through a factory to meet true market demand throws 1960s "push" systems like MRP, ERP, and DRP up in the air. The role of the scheduling and shop floor control modules of these systems becomes unclear. However, deciding at the beginning of the lean implementation to use appropriate modules of these systems—simply as explosion and simulation/planning devices, rather than operating tools—is not a problem. Companies should not delay inevitably pulling the plug on unneeded modules and the accompanying complexity.

THE PERFECT ENGINE

Pull systems are simple and clear, and computer assists have limited, specific applicability in the lean operation. Part of the challenge of pull is identifying the appropriate points where the computer will speed basic operations processes. We think that large computer systems tend at this point in the manufacturing transformation to be somewhat problematic—their cost typically is unjustified, and they become political distractions. Selective application of computers will advance manufacturing speed and integration.

Supply Chain Relationships

Moving from "push" to "pull" systems implies change all the way back in the supply chain, to raw materials and supplier relationships. Developing strong supplier partnerships, with fast, effective, streamlined communications facilitates the move to "pull." Although most organizations understand the big changes supplier partnering have brought to the way we structure and operate supply chains, many companies are struggling with the daily transactions, the actual business of forming workable partnerships.

If your purchasing group is working out of a traditional vendor approach, expect that although this transformation should begin immediately, it will not be completed any time soon. Altering decades of bad practices will require clear demonstration of goodwill and walking the talk.

Cost Accounting System

Standard cost systems based on absorption accounting methods have long proved an obstacle to understanding true costs. Under lean systems, standard cost approaches give way, at least in production centers, to direct product costing that reflects true incurred costs for all the elements of a product, not simply allocated labor, overhead, and materials.

In the area of performance measures, a topic covered in Chapter 8, expect that old methods used to run push systems won't work. Replace outdated efficiency and utilization calculations with metrics that encourage

Preparation for Transformation and Innovation

flow and takt time adherence, rather than overspeed (producing at a rate faster than the actual demand) and overproduction.

Incentive Systems and Management Policies

The remnants of the old industrial revolution are the archaic piecework-type incentive systems that reward associates for more work, regardless of quality or actual demand. These incentive systems encourage behavior that is exactly the opposite of that required for successful transformation to a lean culture. Even some of the profit-sharing and group-incentive systems are disconnected from actual behavior change required to earn an incentive. Most of the time associates in the trenches are as surprised when they are rewarded as when they are not. To be effective and motivating, the lean reward and recognition systems must be simple to understand, with frequent awards—at least once per quarter—and they must provide direct correlation to associates' effort and required behavior change. Lantech and Wiremold have quarterly sharing of gain based on actual improvement in customer satisfaction, lean metrics, and company performance.

In the area of basic management philosophies, lean methods open the organization to a range of "bigger solutions" that include more associates in every process. Narrow traditional measures and methods tend to hold back progress. Expect that your workforce will want to be included in team-based incentive systems, rather than individual bonus or performance plans. As the processes become leaner and more efficient, close-up measures that focus on daily changes—not quarterly or yearly financial measures—become very important to setting direction and immediately reinforce the good results.

As the company moves from valuing machines and systems to valuing people and their creativity, traditional asset evaluation and equipment purchase formulae become less relevant. When kaizen teams create cells and simple machines within them, requests for "monuments"—big capital equipment investments—drop. Compensation and recognition systems need to recognize and reward creativity instead of capital equipment investments, and cooperating teams of people instead of big systems.

THE PERFECT ENGINE

Organizational Structure

A traditional organizational structure, one that is designed for command and control situations and limited, executive-directed communications, is slower and less responsive than lean structures. Further, the alignment of processes and intelligence in the lean world is structured to meet market and customer demand. Unfortunately, traditional hierarchies don't always include customer focus in their structure. Understand, therefore, that whatever traditional organizational elements remain in the company, they will slow response times and customer focus, and they must eventually be transformed to the lean approach illustrated in Figure 4-1.

A Leaner Structure

The first step toward breaking down traditional command and control organizational structures is to think about and deploy production assets around specific products. Each product runs in a dedicated cell. Support services like purchasing, scheduling, engineering, maintenance, and quality may, in the transitional organization, report separately into plant management, although typically such individual projects as new product launches would draw area specialists into mixed teams.

Ultimately simplicity, perfect process, and customer focus become the key concepts in the lean organization. Specifically, to create pull and develop strong and responsive focus in the workforce, the company must deploy all assets—people and machines—along product lines, rather than functions. When teams create product cells, they are drawing all the processing operations into a single focused area; they may even dispatch engineering and design to the floor to work the front end of the product cycle.

Product Focus Teams

After reorganization, however, all functions—as well as operations, scheduling, engineering, maintenance, and accounting—are represented for each specific product. A product focus gives complete line of sight

Preparation for Transformation and Innovation

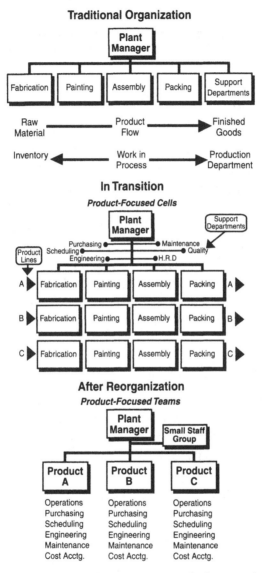

Figure 4-1 Organization charts—before, during, and after transformation.

accountability, from customer order down through shipping and packing, another benefit that builds speed and responsiveness.

The best lean teams, like those at Lantech, stay in touch with their customers by reverse engineering competitors' products, for example, and by visiting customer sites and talking with actual users. It is entirely possible for management and associates to make the customer part of their lean team if the structure is made flexible and responsive.

Workforce Involvement

Moving from the traditional plant structure into the transitional product focus cell and into product-focused teams will have enormous, immediate impact on the lean workforce.

Associates become involved with product concepts long before prototype or volume builds; they are actively working to create new process methods. But management can also use associate involvement as an opportunity to teach financial objectives, to relate a good production day, for instance, to market challenges. At Pella, for example, kaizen teams in the customer service/order administration area have developed in-depth knowledge about distributors and market strategies that no market research firm could provide in less than two years of in-depth study.

True Involvement Builds Trust

Further, the engaged lean workforce will, after a reasonable period of training and transition, find solutions before problems arise. When associates understand that the new structure is offering them job security and new opportunities for daily excitement, they look forward to working and begin to speak more openly about what is really happening in their operation. Lean organizational structures can help build trust and openness.

Union and Contractual Issues

There are some plants, however, that must, before they begin to restructure, deal with various workforce challenges—union contracts, obtuse job descriptions, and complicated pay scales. These are complex issues that have defeated and prolonged many good transformations.

We recommend that management initially address contractual constraints with side agreements, especially where union rules restrict job

Preparation for Transformation and Innovation

classifications and job movement. It is extremely important that job security concerns are addressed up front. Overall, associates and their union representatives need to understand that removing waste and creating new workflows do not mean that associates will be laid off.

There will be, however, greater need for flexibility and training—accelerated change—among the workers. Management needs to recognize from the beginning that some workers will never adapt to change or more decision-making options: Do not expect universal enthusiasm for workforce changes. Deal with every objection and every unstated obstacle off-line with separate discussions and individual problem resolution.

Management's Role: Guidelines for Change Management

Recognize that in addition to creating specific lean organizational structures, understanding, anticipating, and managing continuous change should in itself be a management task. All lean initiatives begin with great enthusiasm and energy, but inevitably they falter, and management must learn to predict and manage the down times, as well as recognize the successes.

Develop a Clear and Compelling Vision

The first step in change management is to establish and communicate a clear vision of change, and to develop achievable, stretch objectives that meet business objectives, but that also energize and encourage the workforce. Sometimes a crisis like Lantech's loss of patents, or Honda's financial problems in the late eighties, creates enough urgency to move an entire organization in the right direction. However, in the absence of a crisis, management's job is even harder, as they must create a clear, concise, and compelling vision based on market realities and the need for enterprise survival over the long haul. Creating acute awareness of the customer, the market, and competition will generate urgency for enlightened leadership, and this awareness provides leadership with a cornerstone communication opportunity. Rallying themes or even annual slogans help drive the change.

THE PERFECT ENGINE

$19.98 Cameras in 1998

For example, Polaroid's marketing slogan, "$19.98 cameras in 1998," kept the focus on cost reduction in order to keep the Scotland plant solvent, viable, and open, against intense pressure to relocate operations to China.

Nissan's "Number One in Customer Satisfaction"*

The entire organization was included in the focus of absolute quality at the time of the Altima launch. Managers participated daily in drive tests and audits; they also called new customers personally and asked for initial quality feedback. No effort was spared and no aspect of the organization was excluded from full participation.

Management Does Make All the Difference

We prefer a systemic change methodology to move organizations out of traditional practices through many difficult transitions into lean methods. Change is an iterative process that uses clear communications, participa-

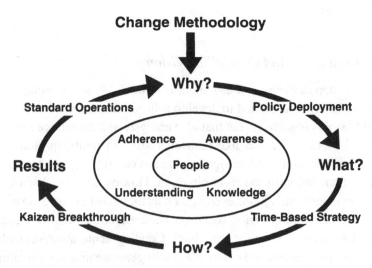

Figure 4-2 Change methodology.

*Quality Index, J. D. Power Associates for Altima, 1992–93.

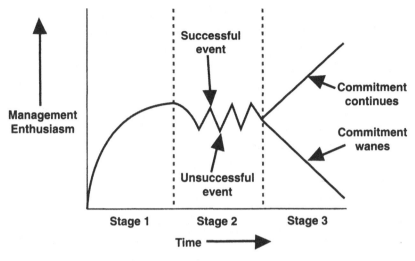

Figure 4-3 Impact of management enthusiasm.

tion by leaders, some failures, and well-recognized successes to maintain momentum.

Even advanced kaizen sites like Pella, Wiremold, Mercedes, and Lantech experience ups and downs, and each of them has had to work hard to move beyond growth plateaus. Sometimes it takes the energy of top management to move the weight of depleted teams, and it is management that can deploy the extra boost to bring teams back on course and on to the next level of process maturity.

Figure 4-3 is a simplified look at the peaks and valleys an organization beginning its lean journey can expect to encounter.

In Stage 1, enthusiasm and blind hope reign; everyone is curious and excited about the new ideas; few associates who have doubts about lean hold themselves back. By Stage 2, however, after the organization has conducted its first few successful and a few not-so-successful events, people are beginning to hesitate and hold back; their enthusiasm is limited and they look for reassurance. This is also the perfect opportunity for all the anchor draggers who stayed quiet during the initial enthusiasm to come out and shoot unfriendly arrows into the backs of the change agents.

Finally, by Stage 3, at the point of full engagement, management's

THE PERFECT ENGINE

commitment becomes the sustaining factor that will carry team efforts beyond their initial excitement and reality checks into sustained improvement drives. Clearly, without management's visible commitment, the effort will die a slow, annoying death either after the first unsuccessful event of Stage 2, or at the beginning of Stage 3. Resistance can be fatal.

Making transformation continually difficult is the reality of the organization changing the people, the processes, and the products. These factors make it additionally difficult to communicate the clear visions, to educate as to the "what and why" of the nature of things. It's much like understanding that one planting and harvest season is not enough—the farmer must repeat the cycle from season to season to yield a productive crop. The same is true of an organization moving through the challenge of transformation.

Managing Resistance: The Role of Management

One of the recognized factors that naturally creep into any challenged organization is resistance to change. Figure 4-4 shows the usual organizational resistance to change in the form of a normal distribution bell curve and the traditional management response pattern, as contrasted with resistance to change and management attention required in a lean transformation.

The traditional environment includes three groups of employees—anchor draggers, the uncommitted mass in the middle, and early adopters. Management attention tends to focus on anchor draggers who prevent the organization from moving ahead. Fortunately, there will always be a few early adopters at the other end of the curve, workers whose blind faith leads them to explore new methods. Anchor draggers are typically very articulate (they have plenty of time on their hands because of inactivity) and make all the noise to grab management attention. Early adopters are apolitical and not concerned about seeking attention. In an environment in which management attention is focused on a few noisemakers, the silent majority does not join in the change and the initiatives suffer.

In an environment of rapid, forced change, however, such as that generated by typical kaizen projects, management must turn its attention to

Preparation for Transformation and Innovation

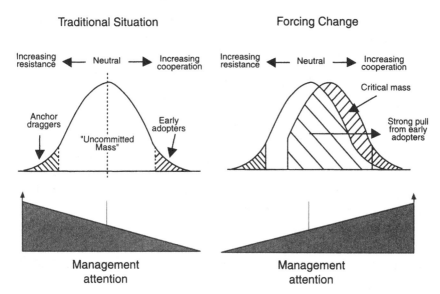

Figure 4-4 Impact of management.

early adopters who will help create critical mass that can tip the balance toward improved cooperation. In essence, where management applies its force is where the fulcrum tips. It is important to understand that although every project has its detractors, by picking the best spot and the best people to begin lean initiatives, and by highlighting and rewarding the change agents, management is making the best use of the company's creative talents. Deflecting attention from detractors will eventually cause them to go away, cease resistance, and perhaps even decide to sign up.

Another Management Advantage: The Kaizen Promotion Office (KPO)

Even as the role of leaders changes to selecting, training, and coaching good contributors, the list of tools and techniques available to management shifts. Not unexpectedly, lean leaders can expect to work with "soft" issues as well as the purely technical ones, such as forming cells, redesigning the process, and developing new metrics. The soft issues are, however, the most difficult and potentially the most draining challenges for almost all managers.

One of the levers that smart lean leaders can engage, however, is the power of an internal change agent. Mel Haught of Pella recognized that he personally could not do the transformation alone. Haught accepted counsel and quickly moved to create a Kaizen Promotion Office to assist.

Funded with the results of productivity gains, and given ample authority and leadership interface, the KPO also provided endless opportunities for learning, honing skills that would be necessary to ultimately lead the change—technically, emotionally, and spiritually.

Executives like Haught and Art Learmonth at Maytag prepared the way for professionals like Brian Giddings, head of Pella's KPO Office, and Ramin Zarrabi, head of Maytag's KPO. These professionals have played big roles in their company's early efforts. They understand where opportunities lie, they can identify and sign up new team members, and they have come to be trusted for their internal technical expertise around products and processes, as well as their thorough grounding in lean principles. In essence they are the facilitators who bring new methods into their plants; they are always two steps ahead of the current process. They will inevitably inherit the next generation of manufacturing management, and for that reason alone it makes perfect sense for companies to rely on their KPO leaders to take big roles in the lean transition.

The Management Assessment: Evaluate Readiness for Lean

Mike Herr, TBM's managing director of European operations, understands the power of the assessment process to design appropriate project plans that minimize pain and fit the company's change potential. Herr spends two to four days with top leaders and selected associates to understand the business and the workforce culture. "The assessment includes interviews," says Mike, "and presentations by management, but what really counts is the walkabout."

Herr likes to walk the factory floor backward from the shipping dock back up to receiving: "Since we have already looked at the product matrix, we have an idea of what should be the potential model line. As we walk the

Preparation for Transformation and Innovation

floor we see the reality of the abstraction." To Herr, the shop floor is an area rich with bits and pieces of key information. "I talk with foremen, and observe the workers' body language—are they open and expressive, and what do they do if the line stops? Is the floor clean and orderly? On a trip through Rover's plant, we saw an unscheduled break—the line stopped and everyone broke out cigarettes. All these things are indicators of potential barriers to change."

Herr tends to look for interruptions in flow—people who aren't working, or materials and machines that aren't moving. "Materials that aren't moving, for example, could be units on the assembly line that nobody is working on—we often see forty people on line loaded with one hundred units, with nothing moving." Mike keys on body language, the way associates answer questions—he notices whether people speak from the numbers.

Dan Sullivan, managing director, TBM West, uses the plant tour walkabout to establish a sense of plant morale. "I make a special effort to communicate one-on-one with shop floor associates. Their receptivity to my attempts—i.e., do they smile, can they answer my questions, will they answer my questions, are they generally engaging with a stranger in their workplace—all of these points can provide a trained transformation professional with the needed data to assess cultural readiness as well as technical level."

There are physical clues all around. Is the plant well lit, are the work centers ergonomically designed? What is the pace of the work—are people working hard and effectively? "We sometimes find people working at a high pace, but not effectively. In the walkabout, we are trying to understand, and to find opportunities for improvement."

The assessment highlights key problems facing a business. At the Mercedes São Paulo truck plant, for example, "The problem," says Herr, "seemed to be inventory. We were talking with Daimler in Europe when they were looking to build a new warehouse. We asked them about the production processes, looked at their true needs, and helped them avoid the construction project they were talking about. It's the difference between using kaizen as an improvement activity and doing kaizen linked to the business strategy." Kaizen experts like Herr want to look beyond immediate symptoms to address the root cause, and he uses the assessment to examine the entire operation.

THE PERFECT ENGINE

TBM Factory Operations Assessment OPERATING CHARACTERISTICS					
COMPANY:				DATE:	
PLANT:					
OBJECTIVE	Past 3 Years (Actuals)			Current Year	
				Goal	YTD
QUALITY					
Finished Product First Pass Yield (%)					
Plant Scrap (% to sales)					
Defects / Unit					
Customer Returns (% to sales)					
Warranty Cost (% to sales)					
COST					
Sales ($) / Associate					
Plant Scrap $					
Rework $					
Labor Efficiency					
Overtime $					
DELIVERY					
On-Time Delivery (%)					
Total Inventory (days on hand)					
• Raw Material					
• Work-in-Process					
• Finished Goods					
Order to Shipment Lead Time					
Manufacturing Lead Time					
SAFETY					
OSHA Recordables (#)					
Medical Cost / Associate					
FINANCIAL					
Net Sales $					
Gross Profit %					
Operating Profit $					
Operating Profit %					
Capital Spending ($)					
Inventory ($)					
Total Head Count					
Salary Head Count					
Hourly Head Count					

Figure 4-5 Simple guidelines followed during an assessment.

The assessment and the walkabout will help lean leaders select the best site for the first project. This initial kaizen experience is important because it is a local opportunity to concentrate everyone's learning in the same area. "You don't have to go to Japan in the beginning for good examples," says Herr. "The problem with trips to plants abroad is that peo-

Preparation for Transformation and Innovation

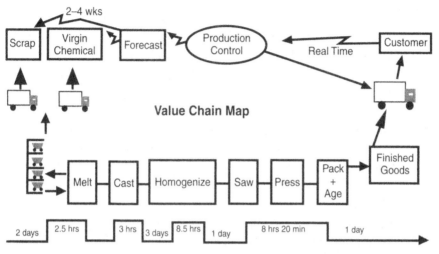

Figure 4-6 A sample value chain map. This one defines aluminum processing.

ple who go to see a good example will see what they want to see—and they will take from it what they want. If they visit Japanese auto plants, and they are in the dishwasher business, it's hard for beginners to make the connections. Instead, we like to see the whole process start right where the product will continue to be produced, on site and local." However, once your company starts the journey and you have gained personal experience, visits to advanced practitioners in Japan or the United States can help refine your own vision.

The assessment interviews include engineering, production control, materials, purchasing, finance, and costing. Information from the interviews and the walkabout is used to construct a first-pass value chain map, looking for disconnects, the flow, and excess inventory or time wasted. "Basically," says Herr, "we are trying to understand total lead time, all the way through the process. The concept goes back to Ford at River Rouge, where ore was transformed to an engine block in three days. Although the individual processes such as melting, stamping, machining, and assembly don't take very long, materials tend to pile up and wait, and that's where the waste is. But Ford understood flow; what he didn't understand was variety."

The value chain map locates opportunities in quality and other time-critical areas. The team then constructs a new value chain map, the "Future State Map," that incorporates possible process changes, that shows

THE PERFECT ENGINE

the impact of cells, for instance, and reduced assembly times. "We spin the web beyond this point," says Herr, "and show not what companies *could* do, but what they have to do. Take Polaroid Scotland, for example."

Polaroid

Polaroid in Scotland makes a commodity product, the black instant camera that is sold everywhere over the counter, a lower-cost consumer item that some planners felt should be moved to a low-labor-cost area such as China. "But," continues Herr, "we argued that this product really is an entertainment product, like the 'Spice Girls' camera, essentially a theme camera that changes packaging as fast as pop culture discards the newest music group." Our assessment showed that this camera needed to be produced close to the market, and we felt that Polaroid needed to produce its instant camera where the design could be quickly morphed to fit new themes. China was too far away to justify the move. The Scotland operation's need for speed and flexibility caused operations to gear up its lean transformation efforts to meet the market demands while maintaining a very attractive lower price. "The assessment," concludes Herr, "allowed us to identify and highlight key strategic data that facilitated the right decision for Polaroid Scotland."

Preparation for Teams

Although self-directed work teams and new product development teams have been working successfully for over ten years, we know that no organization can form and work as a completely team-based unit without thorough training and preparation. The best tool to use for selecting team members and facilitators and predicting their behavior under stress is Personalysis, the powerful personality assessment tool developed by the Houston, Texas, consulting group of the same name.

Personalysis is also an effective tool for selection of key transformation personnel. We often use this tool as an indicator of consulting and influence skills. In addition, the tool can suggest strengths and weaknesses for potential internal change agents. Also, this tool is very effective in understanding and improving the senior management staff's work as a unit, throughout the change.

Preparation for Transformation and Innovation

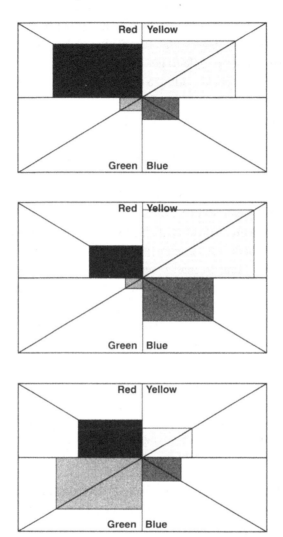

Figure 4-7 This Colorgraph shows the Group Dynamics Profile for TBM.

Personalysis

Starting with a simple, fifteen-minute self-administered questionnaire, Personalysis provides a detailed outline of individual and group personalities. It provides insight into how people think, how they problem-solve, how they deal with others, and how they cope under stress.

THE PERFECT ENGINE

By understanding how people behave and what motivates them, Personalysis enables managers and companies to get the most from their people. It also allows people to get a different perspective on themselves and their coworkers by providing a common, objective language to build and maintain productive relationships.

The test instrument creates colored graphs representing three behavior states—the rational self (the preferred style), the socialized self (learned behavior), and the instinctive self (basic concerns that must be satisfied to feel confident and effective). In each state, four colors represent the individual's degree of influence of four different personality elements—red for "I can do," yellow for "I can influence," blue for "I can choose," and green for "I can control." A team made up of all heavy "red" members is bound for conflict; a predominately yellow team may not get much done, and a blue team member will need time to analyze and make choices, for example.

David Jensen, a Manatech Personalysis consultant, believes that the real strength of the process is helping companies understand personality styles so that teams can take advantage of opportunity, rather than be blocked by soft-side obstacles.

Creativity, the Energy That Fuels Transformation

> Business begins with an idea. And as never before, its growth, stability, and ultimate success depend upon innovation and a continuing flow of imaginative thought.*

Jerry Hirshberg, author of *The Creative Priority* and founding director of Nissan Design International, Inc., the San Diego creative powerhouse that launched a series of market-making, innovative Nissan products, understands the power of creativity unleashed. Hirshberg uses a number of special techniques, Personalysis among them, to bring together teams of unique and incredibly creative individuals under fierce time and cost pressures. It is a challenge that manufacturing companies that have become experts in production, not idea creation, can learn from.

One of the techniques Hirshberg uses to unleash creativity is called

---

*Jerry Hirshberg, "The Creative Priority, Driving Innovative Business in the Real World," *HarperBusiness*, 1998.

Preparation for Transformation and Innovation

disengagement. Sometimes it is important to get distance from a challenge, or to take a break from it and become completely involved in a seemingly unrelated task. Nissan designers developed a tolerance for Hirshberg's surprise events, like trips to the movies, because they came to believe that the rhythm of work, and disengagement, alleviated stress and burnout and unleashed bursts of creativity.

Says Hirshberg, "Our minds will mine any new activity, sifting continually through it for previously unseen connections for bits and parts to fill the nagging void of unfinished, unresolved questions." Hirshberg's approach is not unlike the kaizen masters' urging us to look at nature for solutions to mechanical problems—solutions may eventually come from within, but outside images usually contribute to the process.

Creativity on the shop floor or in the customer service area is not that different from creativity in a vehicle design center. Each area is in a stage of transformation; each process eventually operates under time, cost, and technology pressures, but by brainstorming concepts seen in nature, or disengaging and coming back to the problem at hand, teams will find their creativity is a lever to awaken creative processes so valued in the LeanSigma Transformation.

A Road Map for Institutionalizing Transformation

The kaizen methodology is powerful, and results achieved during kaizen weeks are eye-opening and ultimately translate into customer benefits and the agility to introduce new products rapidly into the market. The LeanSigma Transformation provides a stunning high-performance capability, but the challenge for process managers is to understand the potential and develop strategies to convert productivity into long-term growth and profit improvement. Otherwise, all we have created is an exciting improvement activity that quickly becomes the flavor of the month.

After Kaizen, What's Next?

Rapid change is achieved throughout the value chain using kaizen methodology, but when the organization has moved into a changed state, the next step is to stabilize, or sustain the gains, by focusing on issues that will inevitably surface, primarily:

1. Variability control: With transformation, new variables appear in machines, information, process, and workforce skills. Controlling variability becomes critical to maintaining good process..
2. Improved quality and reliability: Since a LeanSigma Transformation is intolerant of abnormalities, it breaks down at the very hint of quality or reliability problems. Improving reliability and quality becomes fundamental to sustaining the gains.

LeanSigma Transformation Model

TBM uses a proprietary technique called the "LeanSigma Transformation" to overcome the destabilizing effects on the system in the midst of transformation caused by rapid changes of kaizen breakthrough methodology.

LeanSigma, a marriage of lean and six sigma methods, maintains the good results that will be achieved using kaizen. LeanSigma is a stabilizing force, much like a real-time control and problem-solving system that protects, maintains, and builds on sound process. LeanSigma incorporates the best tools of six sigma with a company's conversion to lean manufacturing using the kaizen methodology.

According to Bonnie Smith, a certified master black belt and former director of total quality at York International, now head of TBM's LeanSigma practice, "LeanSigma allows the associates to focus on reducing defects and variability. LeanSigma methods draw on the best principles and tools from both sides of the manufacturing world—quality tools from six sigma, including statistical and analytical tools, and lean tools from the Toyota Production System and kaizen methodology."

Smith sees four phases to LeanSigma, an iterative cycle that recalls the Deming Circle:

1. Measure: LeanSigma will help associates determine how a process is performing, as well as how to measure it. Useful tools include process capability analysis, measurement system analysis, and quality mapping.

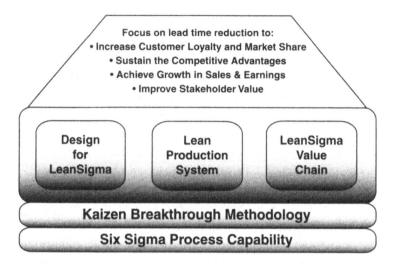

Figure 4-8 LeanSigma Transformation model.

2. Analyze: Analysis should identify the most important causes of defects. Statistical tools like hypothesis testing and simple basic quality tools like Pareto analysis, histograms, and cause-and-effect diagrams are often used during this phase.
3. Improve: The improvement phase of LeanSigma removes causes of defects by using jidoka, poka yoke principles, and process optimization from design of experiments.
4. Control: The job of maintaining improvements under lean methodology uses tools such as standard work, statistical process control, and visual controls.

Six Sigma versus LeanSigma

Although both Six Sigma and LeanSigma techniques focus on the same end result, there is an enormous and significant difference in implementation methods for Six Sigma and for LeanSigma. Kaizen breakthrough methods produce results faster, and according to Smith, "drive culture change a lot better. LeanSigma is a more powerful transformation tool because after kaizen has produced rapid, intense change, LeanSigma maintains gains and reduces variation in the process."

THE PERFECT ENGINE

Further, LeanSigma as a stabilizing force truly allows a transforming organization to focus on very difficult process challenges, the deep-rooted or puzzling questions that cannot be solved simply with kaizen initiatives. At Pella, use of the LeanSigma process, for example, made it possible to track, analyze, and fix difficult technical problems that were causing a high scrap rate and in-process rejects in its paint shop. LeanSigma's root cause analysis and process optimization uncovered in one week the root cause technical issues that would have required long-term solutions.

LeanSigma will resolve issues, according to Smith, that "one cannot easily see. In fact," says Smith, "there is a LeanSigma progression at work here. Processes working at the one, two, or three sigma level can be addressed using logic, intuition, and basic quality tools—histograms, Paretos, and so forth. But then, as your operations are reducing defects and heading toward four or five sigma, you really have to optimize your process, and you will need to use more sophisticated analytical tools to discover root cause and to optimize it. And to really reach Six Sigma, it will be necessary to look at the product and process design. That link," says Smith, "takes you to Design for LeanSigma; for Six Sigma quality levels, the design of product and process must truly be perfect."

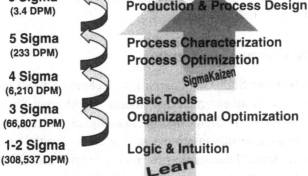

Figure 4-9 LeanSigma progression at work through this process.

Preparation for Transformation and Innovation

It's a matter of levels: "You can only be so good on the floor; if the design is faulty, you won't get there."

Smith reports using LeanSigma tools "everywhere, up front in order administration and customer service, at the back-end in logistics and distribution, and all through production processes."

Transformation Guidelines: Four Simple Steps

Transformation unleashes dozens of questions, and hundreds of possibilities, but we recommend that to stay on track, one keep a few simple guidelines in mind throughout the journey. No matter what tools are chosen from the transformation toolbox, it is important to follow these valuable four steps to keep your company's transformation on the right path.

Step 1. Focus on a Few, Simple, Operational Indices

Typically, we recommend that companies develop seven to ten simple-to-understand operating performance measurements, then implement them systematically area by area. It is important to discard the old measurements—no holdouts—and train operations and financial people in the importance and mechanics of the new measurements.

There are a few simple and standard measures any company can customize. These metrics represent early indicators that must be prepared and put in place in advance of full-blown transformation work because they are the only operational measures that ultimately translate into long-term growth and profitability.

Step 2. Link Transformation to Strategic Business Objectives

Once the performance measurements and targets are established and widely communicated, the focus must shift to the implementation of the LeanSigma Transformation. The successful implementation process must be strategic in focus with aggressive and meaningful improvement objectives. Individual kaizen events must be linked to improve and demonstrate their impact on the entire value chain.

THE PERFECT ENGINE

Quality & Customer Satisfaction
- Customer Satisfaction Index
- Customer Return %
- Average First Pass Yield %
- Defects per Unit

Flexibility & Responsiveness
- Delivery Compliance %
- Total Inventory Days on Hand
- Raw Material Inventory ($)
- WIP Material Inventory ($)
- Finished Goods Inventory ($)
- Average Quoted Lead Time
- Product Development Lead Time

Cost & Productivity Improvements
- Value Added per Associate
- Average Labor Hour per Unit
- Average Material Cost per Unit

Safety & Ergonomics Improvements
- Injuries per 100 Associates
- Medical Cost per 100 Associates

Financial Performance
- Net Sales
- Operating Income as % of Sales
- R&D Cost as % of Sales
- Capital Investments as % of Sales
- Working Capital as % of Sales

Figure 4-10 Key performance measurements.

The next challenge is to make sure that the results achieved during the kaizen event stick without any backsliding. The key here is to transfer ownership of the results and the responsibility for sustaining those results to the supervisor or product line manager in the area.

In the short term, people freed from the line by improvement efforts

Preparation for Transformation and Innovation

can become part of the KPO task force and help improve the area they left, as well as other parts of the business. It's important to prepare the workforce in advance to understand this shift. To make the productivity improvement a win-win process, free the most skilled, experienced, and versatile workers: "Free up the best to help the rest." In the long term, however, companies must grow or insource some of the subcontracted operations to create meaningful jobs.

In the worst case, it may be necessary to reduce total production headcount or hours to achieve real productivity gains. The trick is to accomplish this without any layoffs. Often, the answer is to reduce costly overtime first, or to release temporary workers. Attrition can also be a way to slowly reduce headcount without conducting a forced reduction.

Step 3. Focus Narrow and Deep

The implication for companies implementing lean is clear: They must improve the entire value chain of a specific product or product line. Taking improvements to the extended enterprise not only makes the impact transparent to the customer, it becomes a great learning laboratory for the employees to enhance their understanding of the lean transformation of their own culture, with their own set of constraints. We urge companies to select a simple but visible product line as a model line starting point, and keep improving it regularly—and publicize it—to implement all the aspects of the LeanSigma Transformation, production smoothing, mixed-model production, and Design for LeanSigma. We typically recommend that at least 50 percent of the transformation efforts should be concentrated on the model line during the first six to twelve months, and to a lesser extent thereafter.

In a typical manufacturing environment, managers take a parochial view and want to focus only on the primary value-adding operations like assembly or machining. They conduct kaizen events aggressively across all assembly or machining operations. Although this approach may generate productivity or quality improvements on a localized basis, the global benefits for the customers will not be achieved as quickly. To achieve global benefits and to affect customers' perception of the im-

THE PERFECT ENGINE

provements, focus must shift to the entire value chain of the product, including supply and distribution partners.

Step 4. Visible, Hands-On Management Leadership

We think that each organization's plan of attack, and its results, are different. Some of the methods used are the same, but when creativity and transformation is your goal, you must go beyond accepted solutions to prepare the organization for total change and transformation.

Only management's visible attention and support of big changes will make them "stick," and only senior business leaders have the high ground that allows them to see and manage every element of the change simultaneously—from the supply chain, through the workplace and the workforce, down through metrics and policies.

Design for LeanSigma

Further, in the LeanSigma Transformation, there is one relatively new and very powerful formal method, Design for LeanSigma, that prepares production lines in *advance* of running product to achieve the highest efficiencies and the smoothest flows. Although many operations have imposed older designs of lines and equipment on new or revised products, designing the best process to enable production to achieve its overall goals is unquestionably the best route. Design for LeanSigma is also used successfully to design new products with unleashed creativity and LeanSigma principles. It incorporates the best creative brainstorming work with sound industrial design and lean practices to create new products and processes using solid ergonomic practices.

The transformation journey is difficult and requires constant attention, including confronting the technical challenges of creating flow and pull where they remain to be done. Further, much effort will be devoted to managing the culture as the organization moves spiritually and physically, and measures its performance in different ways. All LeanSigma tools and techniques will be required to initiate change, to create new enterprise focus, to eliminate waste, and, ultimately, to positively affect

Preparation for Transformation and Innovation

customers. All stakeholders, however, will take home a piece of the victory.

Transformation is an enormous task, and one that many managers find themselves unprepared to begin, but all movement spells opportunity, and change is almost always good. There is no risk in beginning.

CHAPTER 5
Lean Production System

The Kaizen Breakthrough Experience

Every Monday in the United States, Europe, and Latin America, in any number of foreign languages and all sorts of businesses, teams of individuals gather to improve their companies in the space of one week or less. They are driven by a sense of urgency even though they work for some of the most successful companies in the world. They are committed to teamwork because they have seen the incredible changes the Kaizen Breakthrough Methodology can bring.

A Kaizen Breakthrough is a team-based, up-close focus on solving a specific issue in a company, with set goals and parameters. First, a cross-functional team is handpicked from various departments inside, or outside, the company. Then specific goals are set for the team: productivity gains, floor-space reductions, ergonomic and safety features added. Within a week—or just two to three days, as in the case of *point kaizens*—the results of the kaizen are presented to upper management.

It is a core tenet of the Kaizen Breakthrough that the team has complete support from management and has all the resources they might need to complete their task. At the same time, the team is expected to put creativity before capital as they pool their talents to create a better work environment and increase the company's profitability.

TBM conducts public kaizens, inviting complete strangers to participate in solving real-life issues at a client company. (Following are journal entries from one recent participant.) We share this to show how kaizen can become a way of life—a form of self-reliance that empowers hourly associates and supervisors alike.

On Monday morning we arrived at the Chattanooga, Tennessee, Marriott. There were more than fifty of us, mostly senior executives from different industries all over the country. Some people had traveled all night from Brazil and others from Mexico and England. For most of us, it was our first full dose of the system that had grown from the roots of Henry Ford's River Rouge plant, then Taiichi Ohno's work at Toyota and twenty years of successive customization for North American, European, and South American industries. Of course we had all heard something of just-in-time and kanban, but we had a feeling that this LeanSigma Transformation experience was bigger than these singular pieces of systems that had begun the new industrial revolution. We had all come to learn the secrets of this revolutionary phenomenon with a wide range of personal expectations.

Day One: Learning the Basics

Our first kaizen experience started surprisingly smoothly. We met in the auditorium at 1:00 P.M. for intensive classroom training. We were hoping we would learn the basics of lean manufacturing, if there were such things, and it seemed that we were not going to be disappointed.

Our instructors, Sam Swoyer and Bob Wenning, both TBM consultants, and now our guides throughout this five-day program, wasted no time in putting us to work. Sam explained the challenge—after this day and a half of classroom instruction—we would be assigned to teams and sent off to work areas on the shop floor, each of us armed with shop floor kaizen tools such as a bright yellow plastic stopwatch, our observation and timing charts, and lots of paper and pencils. There would be no phone breaks, no distracting cell phones and no intellectual debates about hypothetical arguments, no excursions to local restaurants, and no down-time.

THE PERFECT ENGINE

We started with the basics—an overview of time-based strategies and the LeanSigma Transformation—but soon found ourselves deep in the charts and graphs that we would use to describe the current production process, and to imagine the new ones. Training wrapped up with an introduction to our host plant by its senior management, at around 8:00 P.M.

Day Two: Initial Observations, Learning to See

The next day we were back in the classroom at 7:00 A.M. for more training and instructions for the afternoon's trip to the plant. After lunch we all boarded buses for Cleveland, Tennessee, Maytag's Cooking Products plant. The plant we were set to work in was an established and profitable producer of home cooking appliances. It was large and the product flow was long and complex, with a full range of operations and equipment, from long assembly lines, to scattered feeder departments like sheet metal fabrication, stamping, milling, drilling, welding, and subassembly.

We broke up into six teams identified by the color of our caps, Red Hats, Brown Hats, Green Hats, etc. It was clear when we arrived at Cleveland's main plant, an enormous complex of manufacturing buildings rising from the Tennessee hills, why the eleven-member teams wore quite visible caps. Without an inside guide and our brown kaizen team hats, our team would be lost in the maze of noisy machinery and raw material storage in less time than it took to expedite a lost oven door panel. Even finding the bathrooms was a challenge—one team member wanted to drop sticky labels along the yellow stripes leading to the giant washrooms.

Our first team exercise was an official guided tour of the plant put together by local KPO (Kaizen Promotion Office) leaders, stopping at several bottleneck areas and worksites for other teams. Finally, we arrived at our worksite, a portion of the Jenn-Air Downdraft Cook Top assembly line. After observing the ten operators in our focus area and studying the flow of materials—hundreds of fabricated and purchased parts as well as sub-

Lean Production System

assemblies—the major pieces of the operation were starting to fit. A quick sketch revealed three or four problematic operations. Several departments fed the assembly line, and a final long lurching line of nearly finished products waited for various shortages and finishing operations. Soon after that our consultants dispatched us to the floor in teams of two to observe and measure individual operators' tasks and compare them against the rhythm of the line.

Learning to See by Following the Numbers

It's not unusual for teams to find themselves hunkered down with a problem that defies definition—lateness, for example, can be a code word for "unhappy customers," or out of control scheduling processes, or even a constant expedite mode. Sometimes the only way to truly define a problem is to cut away the symptoms and let the numbers lead you, a key teaching of Dorian Shainen, Shewhart Award winner and quality pioneer. Shainen believed, and the Brown Team discovered, that uncovering key numbers that described the process—hundreds of SKUs—for instance, explained the huge problem dealing with variety. Other numbers gathered quickly by team observers armed with clipboards and stopwatches showed suspiciously uneven operator cycle times against the fixed pace of the line. In one area, workers kept a frenzied pace, while other operators worked at a leisurely pace, appearing almost bored. "Hurry up and wait" was the work routine. However, once we plotted our observations on a Cycle Time/Takt Time Chart, it was clear that our cell was operating to a broken rhythm.

But how was one team to fix such big problems, achieve seemingly impossible objectives—25 percent productivity, 50 percent space, and 40 percent WIP inventory improvements—in two and one-half days on the shop floor? The task seemed overwhelming, and we weren't in agreement even about where to start. We returned to the hotel that night at 8:00 P.M., tired, confused, and discouraged.

THE PERFECT ENGINE

Day Three: Dirty Hands and Dusty Sneakers

Another day of observation, walking, walking, walking, and arguing and sharing of solutions, ended in complete frustration and unpleasant team dynamics. No one had promised that the experience would be a day at Disney, but none of us was comfortable with the seeming lack of resolution.

Sometimes the answer does not lie in our heads, or in the stacks of "explanatory" printouts. Sometimes the solutions are on the floor, locked inside the quietest shop associate. Three of us vowed to revisit the floor that morning, hoping to find just a few answers out there. We were not disappointed.

We knew material movement and loading and unloading heavy subassemblies was a problem. We observed that the uneven flow of components to the line, as well as awkward part presentation, coupled with lack of proper tools and standard methods, disrupted the flow quite frequently. The wiring operation, to connect the control panel logic to the guts of the unit, was particularly tedious and an ergonomic nightmare. Two operators had to kneel and bend their bodies at an uncomfortable angle to perform various wiring steps.

Wednesday morning saw an empty War Room as subteams attacked the Downdraft line to implement simple but effective solutions. We knew we could not fix every big and small problem in the three days. But every solution starts with a few good fixes, and that's where we started our work. Wheeled tables, scissor lifts, and gravity feeds, mocked up and assembled from scrounged parts, easily fixed the sore backs and overused arms of shop floor associates. Moving empty tall racks out of the line, and cutting down the length of the line by almost 50 percent, consolidated the line for efficiency and created line-of-sight manufacturing.

Design of Experiments

The kicker was the multitude of solutions to the ergonomic issues of the wiring operation. Along with my subteam member and a maintenance mechanic, we tried five different experiments, each progressively improving the operation, but with limited overall success. Each *solution improved the ergonomics but was not practical or justifiable. We returned from the plant at 10:00 P.M., very tired but very excited.*

Day Four: The Breakthrough

Thursday our team arrived at the plant at 7:00 A.M.; all of us had dreamed many more solutions to some of yesterday's problems, but first we wanted to see the progress made to the physical changes we had asked of the maintenance team. To our pleasant surprise, the new design of the line was almost complete. Total length of the line was reduced by half and now it required only thirty-one instead of forty-four operators. We spent most of the morning fine-tuning improvements, changing some of the areas as many as seven times. Our team's spirits were high and most of the operators were generally pleased with the improvements in their work areas and the new work methods.

The Maintenance Miracle Worker

At about 11:00 A.M. our friend the maintenance mechanic arrived late and tired but in very high spirits. He told us that last night he could not sleep and kept on working on the wiring ergonomic issue in his garage. He took off a spring from his boat, and some other parts from around the house, and built a sleek and robust fixture that solved the wiring problem. This fixture would ride on the assembly belt and carried the entire unit on it. However, when it arrived at the wiring station the operator could tilt up the unit at forty-five degrees, with a simple touch of a lever, and complete all wiring steps in comfortable standing position. Next, the operator could easily tilt the unit back down again for normal operation at the next station. The operators loved it. The maintenance manager estimated that he could duplicate "the new carrier" with less than two hundred dollar in materials, or approximately six thousand dollars for all the fixtures needed in the line.

During the afternoon's actual production and retiming run, we also discovered that with the ergonomic improvements, we required

THE PERFECT ENGINE

only one person for wiring, instead of two. That was certainly an unexpected surprise.

Day Five: Documentation, Presentation, and Celebration

Preparation to celebrate the results of a kaizen event involves hard work to complete all the standard work documentation, prepare a script, and, for some team members, even more work to deliver the message to management and the kaizen masters. Preparation is 90 percent of the exercise—each team member's explanation of results and implemented solutions takes less than two minutes, but rehearsals seem endless. It's difficult presenting to the masters, giants whose lives are filled with daily kaizen challenges and seemingly simple solutions.

But we get through it, the Brown Team unscathed and well represented by good work in a critical area. We were all amazed at the speed of change and the number of improvements that actually got implemented during the five days. For me it was truly a breakthrough experience.

When Taiichi Ohno visited the Ford Motors plant at River Rouge in the early fifties, he was truly humbled. Ford Motors' quality and productivity were several times better than Toyota's. The operational lead time, or total elapsed time for converting raw materials into a Model T, was only three days. Mr. Ohno worked diligently over the next twenty years to develop a version of Ford's miracle that came to be known as the Toyota Production System. The Lean Production System, developed by TBM Consulting Group during the eighties and the nineties along with the Shingijutsu Company, is based on Toyota Production System principles, expanded for application to all industries and cultures. Experienced users find tools inside the Lean Production System to identify and remove all forms of waste—from wasted operator motion to wasted process time and poorly used raw materials or unneeded investments. As an operations tool that has been perfected on the shop floor, the Lean Production System produces results quickly when applied with discipline and energy.

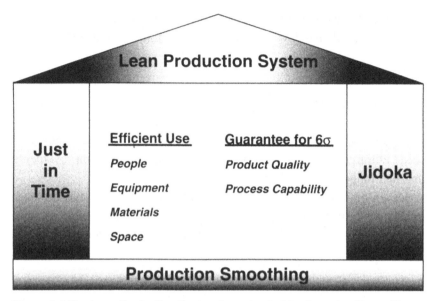

Figure 5-1 The Lean Production System is supported by the twin pillars of just-in-time and jidoka, resting on a foundation of production smoothing.

This diagram shows a simple but comprehensive application of many of the concepts of the Lean Production System applied in discreet manufacturing, from planning, machining, subassembly, and assembly to the integration of suppliers through kanban pull signals.

Three Components of the Lean Production System

Let's take a closer look at the three main components of this powerful system:

1. Just-in-time improves customer service, reduces lead time, and increases utilization of critical resources, such as people, machines, materials, and space, by eliminating the system waste.
2. Jidoka focuses on making the process autonomous so that it can run unattended and automatically stops at the first defect or abnormality. Jidoka includes many quality tools and techniques, like poka yoke (mistake-proofing) and statistical process control.
3. Production Smoothing is the adaptation of scheduling algorithms to

THE PERFECT ENGINE

insulate manufacturing operations from the daily fluctuations of customer demand, while ensuring the customer is satisfied.

Just-in-Time

When companies begin the Lean Production System journey, they frequently focus on "easy hits" for rapid improvement—shorter lead times and work-in-process inventory reductions—through one-piece flow and waste elimination. It's a great place to find balance sheet improvements that fuel longer-term initiatives.

Just-in-time is a manufacturing system that produces what the customers want, in the quantity they want, when they want it, while concentrating on the minimum use of raw materials, equipment, labor, and space.

Principles

Time is a precious operating commodity, just as valuable as steel and lumber, but more difficult to buy. By linking production to the actual customer demand and reducing manufacturing lead time—the time to convert raw materials into finished products—to a bare minimum, a producer can achieve best quality, cost, and delivery advantage. When Pella, for instance, cut lead times the company found that just-in-time or lean methods offered the Iowa competitor new market dominance opportunities. The distributors and end customers were delighted. Successful implementation of lean requires strict adherence to the three basic principles:

- Creation of flow production, or one-piece flow
- Adherence to a pace that equals takt time, or the beat of customer demand
- Discipline to produce only at the pull of the customer

Flow Production

One-piece flow is an enormous departure from traditional batch or "push" methods and one of the most challenging Lean Production System concepts. It produces epiphanies among converts, who quickly discover system flaws

Lean Production System

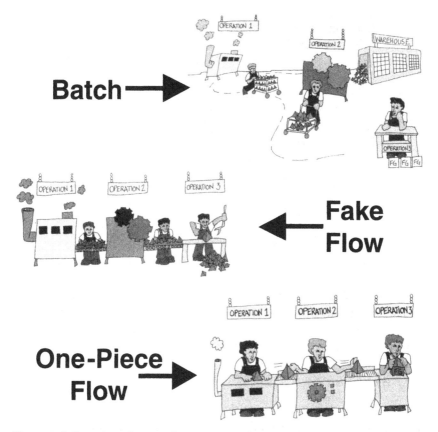

Figure 5-2 One-piece flow production versus batch and "fake flow."

when they are forced to complete one entire unit at a time, in one place, without any detours or backtracking. Black & Decker discovered the power of one-piece flow when it disassembled its iron assembly lines and changed them to multiple one-piece-flow cells, with a fourfold increase in productivity.

By pacing the operations to takt time and developing alignment with the actual market demand, companies save time and increase customer satisfaction. Takt time, the beat of an operation, is determined by the actual rate of demand from customers. It is calculated by dividing available manufacturing time in a period by the actual demand during that period. If an operation runs only one shift a day, it has 450 minutes of available time (after deducting two ten-minute breaks and ten minutes cleanup

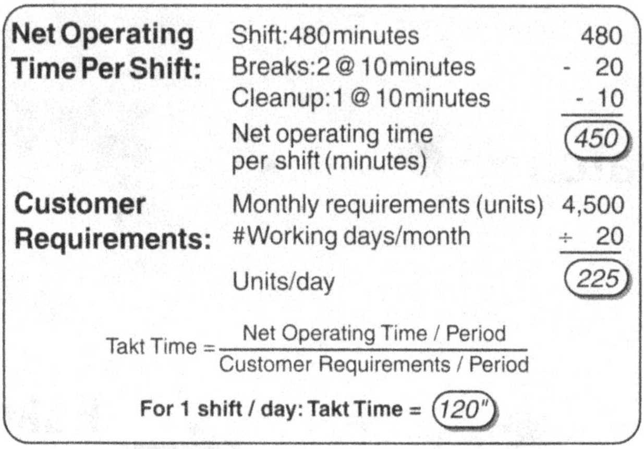

THE PERFECT ENGINE

Net Operating Time Per Shift:	Shift: 480 minutes	480
	Breaks: 2 @ 10 minutes	- 20
	Cleanup: 1 @ 10 minutes	- 10
	Net operating time per shift (minutes)	(450)
Customer Requirements:	Monthly requirements (units)	4,500
	# Working days/month	÷ 20
	Units/day	(225)

$$\text{Takt Time} = \frac{\text{Net Operating Time / Period}}{\text{Customer Requirements / Period}}$$

For 1 shift / day: Takt Time = (120")

Figure 5-3 Example of takt time calculation chart.

time from an eight-hour work shift). If daily demand is 225 pieces, then the takt time of the operation will be 120 seconds. In other words this operation should produce one product every 120 seconds.

PULL SYSTEM

When a system responds only to the customer's *pull,* it requires that each component of the flow manufacturing system responds only to an actual demand signal from the successive operation. The system virtually links the actual customer's buy signal through distribution, various manufacturing operations, all the way back to the supplier of components and raw materials. One purpose of the pull system is to shut down an operation as soon as an abnormality occurs. The abnormality may be due to disruption in demand, machine breakdown, or a quality problem. In any case, the pull system's discipline will surface the abnormality for immediate corrective action and root cause analysis to make the operation more robust.

LINE-OF-SIGHT MANAGEMENT

Lean factories are smaller and cleaner, with better line-of-sight management over fewer, simpler machines and better ergonomics. While a tradi-

Lean Production System

tional plant follows a functional layout—all grinders, presses, and assembly operations separated into their own departments—the lean layout includes product-dedicated cells and short lines with appropriately sized machines. There are minimum operators, minimum material lifting, and the least movement possible from one area to another.

When big volume shifts require adjustments, it is easier to add or subtract workers in a lean environment and to create a new standard work routine than to ramp up (or down) a hardwired process. Moreover, because feeder operations and subassembly lines in traditional manufacturing plants may be located far from final assembly, it is almost impossible to really understand and calculate the full impact of a major volume shift. In traditional factories line-of-sight is not always clear and sometimes it takes months to accumulate the net results of volume shifts. By the time the next shift happens confusion reigns, resulting in frequent stock-outs or fire sales.

CELLULAR MANUFACTURING MAKES ERGONOMIC SENSE

Operators working in a cellular environment find work less tiring and more rhythmic. Because simple mechanical assists—rollers, scissor lifts, and tables and tools placed at the right height—require less out-of-range body movement, and innovative tool presentation eliminates wasted operator movement, the work routines are more predictable.

Jidoka Is Autonomation

The literal translation of jidoka is "automation with a human touch." The operational application of the concept means separating machine work from human work, and giving machines the capability to detect and immediately respond to production abnormalities. Taiichi Ohno of Toyota coined the English word "autonomation" to describe jidoka.

Jidoka seeks to simultaneously meet the customer's need for the highest possible product quality and achieve the most cost-effective manufacturing process. *No decision about layout, or machinery, or worker training and movement can be made independent of each other factor that makes up the production system.* Although traditional pro-

THE PERFECT ENGINE

duction thinking might approve of making capital equipment purchase decisions based simply on time-to-recover the original investment, jidoka looks at how workers will work with machines, and how the machines can be made to work independently to perform simple, repetitive functions.

USE THE PROCESS TO PREVENT DEFECTS

Further, jidoka machine design builds defect-prevention into the equipment because tools and fixtures, as well as processing operations, are designed to detect problems and shut down when they occur. A punch press can be designed, for example, to detect incorrect placement of sheet metal on the table. The machine shuts down when tolerances beyond the pattern layout are detected, long before human eyes would notice and respond to out-of-tolerance run conditions.

Production Smoothing

Production smoothing means adapting production rates in accordance with customer demand, despite variations in volume and product mix.

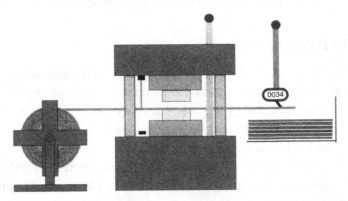

Figure 5-4 In this illustration of jidoka, the press is equipped with a light sensor to detect the presence of sheet steel and a second sensor, above the output stack, senses when the tray is full.

Lean Production System

This is accomplished by calculating the takt time based on an average demand during the demand horizon, typically one month. Every item in daily demand is built every day. If the total demand is less than a day's set production rate, production continues on the most common models. The overflow is then ready to be used when demand is high. Less popular or low-run models are only made to customer demand.

The goal is to maintain a consistent production rate and to use inventory to buffer the upward and downward shifts in demand (see Figure 5-5). For model mix variation, a daily production schedule is developed for each model, with an ultimate goal of producing every model every day, to improve responsiveness and consistency in the process. Ultimately, production smoothing of volume and mix helps plants respond daily, just-in-time to customer demand, while keeping costs down.

Lean Production System Model Line

Figure 5-6 shows how all these concepts of Lean Production come together to create a line-of-sight production process that is integrated with the front-end planning systems and linked to actual customer demand. Visualize a pull system: When up-front planning has erected clear boundaries in the process, a customer's pull can touch the end of the assembly line and spark the system into motion.

Kaizen Breakthrough

Implementing a LeanSigma Transformation that incorporates all the best Lean Production System practices while maintaining shop floor associates' active involvement is a big challenge for most transitioning industries. That's where kaizen breakthrough, a powerful and energizing structured change methodology, comes in.

Expert kaizen practitioners like Wiremold, Pella, Maytag, Mercedes, and Lantech use weekly kaizen events to transform operations and prepare for new products. Practitioners like Black & Decker, Pratt & Whitney, and Vermeer are using kaizen to improve profits and speed new

Level Loading for Fluctuations in Volume

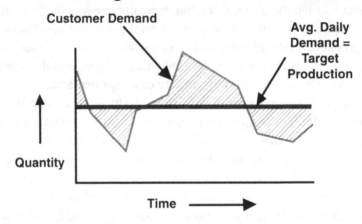

Level Loading for Fluctuations in Mix

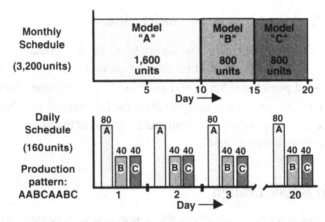

Figure 5-5 To control fluctuations in production, we assess monthly demand by product and create a daily schedule that provides every product in sync with the customer's daily volume demands.

product launches. Basically, kaizen is a team-based method that can be applied to any process to make it better.

Kaizen, the combination of Japanese symbols for change and good, commonly translated as "change for the better," seeks fast and practical

Lean Production System

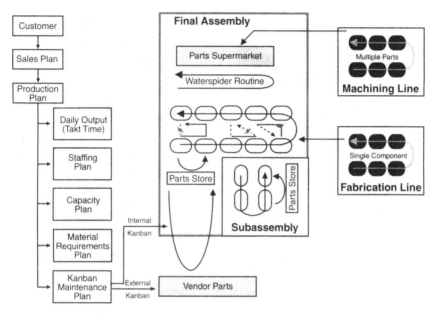

Figure 5-6 Lean Production System model line.

solutions to everyday challenges. Kaizen puts process intelligence and decision-making responsibility squarely in the hands of shop floor experts, supported by actual observation of facts. This is a big change for many operations supported by thick layers of management and support professionals—engineers, equipment experts, materials and scheduling planners—who tend to make all the operating decisions.

Kaizen Moves the Focus to Operations

When pioneers first discovered just-in-time manufacturing, they realized that they would need a very fast and powerful method to bring the full force of Lean Production System experience to the workforce. Learning the system—how to set up cells, what to do with material flows, how to set the work pace, and other softer issues—leads most practitioners to the conclusion that the Lean Production System is best learned on the floor, not in classrooms. That is why we spend little time in the class-

Figure 5-7 Japanese characters meaning Continuous Improvement.

room explaining, and more time *doing,* in kaizen events. A typical kaizen event is successful because it is team-based, has a sense of urgency, and all supporting resources are made available at the team's disposal.

The power of the five-day kaizen event is the energy and intelligence unleashed on the shop floor. Not all benefits of good kaizen activities last, however; it is easy for organizations to slip back to some of their earlier, more traditional practices. In Chapter 8, we talk about metrics that reinforce the new process. Making sure the new way of doing work is the same today as it will be in two months is the job of creating and adhering to Standard Operations.

Standard Operations

In a lean environment an often overlooked and misunderstood tool of continuous improvement is the correct implementation and use of Standard Operations, or standard work. Standard work is the documentation of each action required to complete a specified task. Standard work

Lean Production System

> **What Is the Kaizen Breakthrough Methodology?**
>
> - A cross-functional-team-based process for rapid improvement with:
>   - Bias for action
>   - Creativity before capital
>   - Focus on results
>
> - Focus on physical transformation
>   - Learn by doing
>   - Overcome resistance
>   - Instill change culture
>
> ***Promote Rapid Change Through Involvement!***

Figure 5-8 Description of kaizen methodology.

should always be displayed at an operator's station and updated often. In the multinational environment in which most companies find themselves, using photographs or line drawings to illustrate standard work is an excellent idea. When we research the companies that have not sustained the gains of kaizen, we usually find that the reason is the lack of standard operations.

We also find that the proper use and the power of standard operations have not been clearly understood. Sometimes companies have not documented the process properly or managers do not know exactly how a job is performed from one cycle to the next. There may be too much variability from one cycle to the next, or from one operator to the next, reflecting an absence of standards.

Standard Operations—Unglamorous, but Necessary

Implementation of standard work is time consuming and can be intimidating. It's less exciting than the initial kaizen event, but equally important.

It is also important to note that standard work is not work standards, in

THE PERFECT ENGINE

the classic industrial engineering definition. Work standards are developed by industrial engineers who may or may not have observed the process and determined elemental task times based on averages of "known" work elements. These work standards include the seven wastes, as well as "personal and fatigue" allowances. By putting in factors for "P&F," the company is accepting that it is okay to have waste in the process and there is nothing that can be done about it. The company believes that it has therefore determined the "best" methods that can be developed.

On the contrary, standard work is developed by the supervisor and shop floor personnel themselves, reflecting actual observations and input from the operator doing the task. The concept in Standard Operations is to reflect the work under normal conditions. Since we know that we can never be perfect, and Mr. Murphy does appear from time to time, Standard Operations shows us all the process abnormalities for immediate countermeasures. We can define Standard Operations as the best combination of machines and people working together to produce a product or provide a service at a particular point in time.

Standard work solidifies the improvements made during kaizen events and helps the organization minimize the variability of a given process, thus improving reliability. Sometimes variations develop from one worker to another; sometimes variations happen when equipment tooling has been designed to allow too many work methods. And sometimes variations in standard work creep in unnoticed.

Standard work in essence becomes the documented benchmark; improvements are always possible, but they must be deliberate and not accidental changes in the way the process operates.

Once kaizen team members have redesigned an operation, three standard work elements reinforce continued good practice—documentation of each operation, displays of the documentation, and operator training. Based on the principles of the Lean Production System, there are three elements of standard work that need to be developed and documented for ongoing, consistent performance:

1. Takt time/cycle times chart
2. Work sequence
3. Standard work in process

Lean Production System

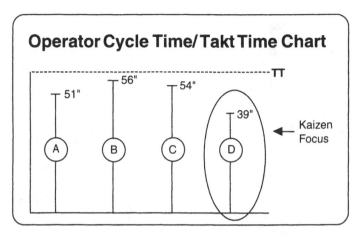

Figure 5-9 In this example, four operators are working on a process with a sixty-second takt time. Each operator's actual cycle is much shorter than takt, however, with operator D needing only thirty-nine seconds to finish his job. We can focus a kaizen on this issue and possibly reduce the number of operators.

Takt Time/Cycle Time Chart

Takt time is the rate at which the producer must produce a product or perform a service to satisfy the actual customer demand. Cycle time, on the other hand, is the time an operator takes to complete one cycle of operation. By plotting the actual observed cycle times for all the operators in a cell against the takt time, we can see the opportunities available for waste elimination.

In addition, by adding all the cycle times and dividing the sum with takt time, we can determine the ideal number of operators required in a cell. Of course, as we improve the cycle times by eliminating waste, this number changes as well. This chart is a very powerful tool in guiding the team members as to where they should focus their energies in making overall improvements to the process.

Work Sequence

Work sequence is the steps an operator is assigned to complete in one cycle. The Standard Work Layout sheet (Figure 5-11) is used to illustrate

THE PERFECT ENGINE

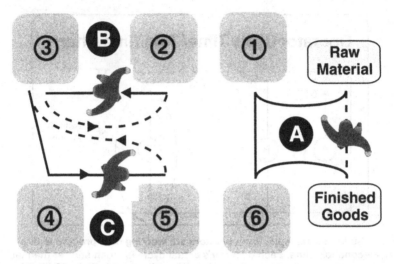

Figure 5-10 Chart showing work sequence.

the proper work sequence for each operator. It should be noted that work sequence does not necessarily follow the material flow in the cell.

A Standard Work Combination sheet also describes the standard time elements required to perform the work. A graphic display of these elements against the takt time gives the operators a quick reference of the work requirements and expectations.

Both of these tools, the Standard Work Layout sheet and standard work illustrations, are essential for operator training and reference and must be displayed in the cell at a prominent and visible location. In addition there is a Process Capacity Table (Figure 5-12), a planning tool for the supervisor to calculate the maximum capacity of the cell based on critical machine capacity—after adjusting for predictable nonproduction interruptions like tool sharpening.

Standard Work in Process

Standard work in process is the minimum amount of work in process (WIP) required to complete the work sequence on demand, and to ensure one-piece flow to takt time. Standard work in process describes and limits

Lean Production System

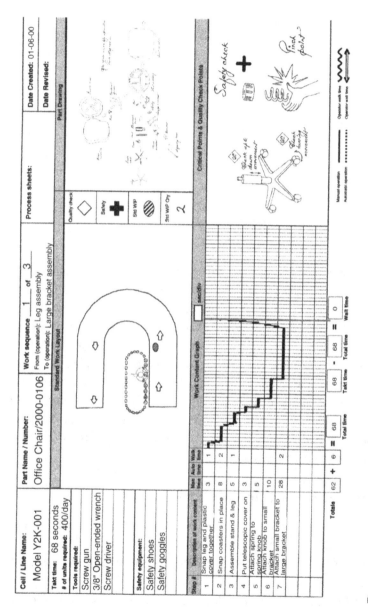

Figure 5-11 This chart reveals the correct layout of the work cell, a complete parts inventory showing everything required to build the product, and each step the operator takes within a cycle. A series of quality checks is also illustrated in line drawings.

129

THE PERFECT ENGINE

Net Operating Time (I) = 27,600"   Daily Requirement = 690

Date: 6/4/92 | Part Number: 1-12930 | Page 1 of 1
Supervisor: Jones | Part Name: Shaft | Max Output / Day: 762

Step #	Process Description	Mach #	Walk Time	Base Time			Tool Change Time			Total Time	Total Capacity	Comments
				Manual Time A	Auto Time B	Mach CT C=A+B	# Pcs / Change D	Time to Change E	Time / Piece F=E/D	G=C+F	H=I/G	
1	Pick up RM		2	2		2						
2	Lathe turning	L400	2	6	26	32	200	50	0.25	32.25	855	
3	Grinding	G220	2	6	30	36	50	10	0.2	36.2	762	
4	Slot keyway	MC110	2	5	18	23	100	60	0.6	23.6	1,169	
5	Inspect		2	7		7						
6	Put down FG		2	2		2						

Figure 5-12 Process capacity table.

the location and quantity of material—subassemblies, components, raw material, and semifinished units—that should be maintained at all times in the operation. Typically, there is one piece of standard WIP at every automatic operation and one piece at each hand-off point between operators. Additionally, if there is a process within the work sequence that takes longer than takt time to complete (i.e., drying or curing), the number of pieces of standard WIP is determined by dividing the process time by takt time.

The measure of standard WIP (SWIP) is typically a definite number of pieces and not piles of inventory. SWIP measured in hours or days-on-hand indicates a bad system and interrupted flow with buffers that hide defects and make the process unreliable.

Managing for Daily Improvement

Standard work is an unglamorous but most important element of lean that maintains the progress kaizens have created. Taiichi Ohno underlined his belief in this foundation concept, "Where there is no standard, there can be no kaizen."

Standard Operations and their documentation form the basis for improvement. Once each of the operations has been observed and the Standard Operations have been posted, the real business of improvement begins.

Lean Production System

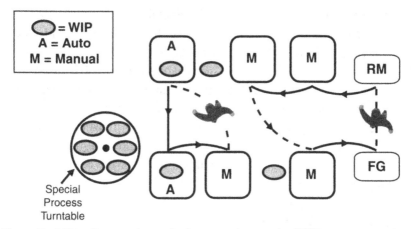

Figure 5-13 This diagram shows the locations for standard WIP in a process: between each operator's hand-off points, at each automatic machine, and where any time-based processing element, such as curing or drying, takes place.

We recommend that the supervisor review each set of Standard Operations with each operator at least once per month. The purpose is to be sure that the operator is following standard work and to ensure that any improvements, even if they represent only a few seconds of waste elimination, are reflected in the documentation. Standard Operations should not be allowed to stagnate—by definition, they are dynamic and must be carefully handled.

Visual displays of Standard Operations on the floor reinforce adherence to the process and serve as a training tool for new operators. Illustrations are invaluable when assessing current conditions and making daily improvements based on identified variances from the standard.

In the Lean Production System, standard work is one of the most important means of maintaining the gains from a kaizen activity and continuing to improve the process beyond the kaizen event. Standard Operations are at the soul of a LeanSigma Transformation.

LeanSigma: Advanced Quality Tools

When companies such as Mercedes, Pella, Maytag, and hundreds of others began their lean journeys, they turned to the Lean Production System

THE PERFECT ENGINE

and kaizen to get moving. As they reduced unneeded inventory, like lowering a water line to expose the rocks, they discovered the need for even more advanced methods of uncovering the root cause of abnormalities.

For these companies, the LeanSigma Transformation has taken them from an intuitive level, in which trained operators and engineers can begin to see and fix issues, to the more complex. At the higher level, where issues are not obvious to the casual observer, we use statistical tools to uncover abnormalities. The journey can take an entire enterprise from the first steps to the most advanced—from an initial kaizen creating the first working cell, to deploying LeanSigma black belts to discover the causal effects of raw materials and processing on finished quality. Each level of transformation takes place in a cross-functional team environment, with collaboration and data collection and analysis driving cultural change.

Champions, Black Belts, Green Belts

This is the more advanced piece of the transformation, in which LeanSigma champions, black belts, and green belts, armed with statistical tools, are deployed throughout a business to root out variation and abnormality. To successfully launch this stage of LeanSigma, we suggest that flow be first established either on a model line or companywide.

Once flow is established, our best practice is to first raise awareness among the top executives who will oversee the LeanSigma effort with a week of champion training. For executives, it's key that they not only learn the tools and techniques of LeanSigma in the five-day training course, but that they also develop the vision and deployment strategy for this powerful program. In champion training, executives assess their current metrics and then focus on defining the business objectives that will drive the projects. As a team, executives may go on to hand-pick their first group of black belts, ensuring that the front-line change agents get the support and resources they require. They will know what training is involved to become a green belt and know the difference between the two types of projects.

Black belts tend to be picked from the most enthusiastic, educated leaders in the company because each black belt must become an effective change agent. Following four weeks of training, culminating in a

Lean Production System

project that must show substantial financial benefit to the company, the black belt is ready to initiate projects and mentor green belts.

The green belt typically has two weeks of training, plus ongoing mentoring. Green belts are most often selected from support groups, such as quality, engineering, and supervisory groups, but they can also be operators. Green belts devote a minority percentage of their time to individual projects, but they are recognized as experts and up-and-coming leaders. In each course of training—champion, black belt, and green belt—we focus on transferring our knowledge to others because we want our client/partners to take control of their own destinies.

Getting Started

Many companies want to see the instant benefits of LeanSigma, even before they roll out their training. We prefer that an entire company—and enterprise—be involved in the effort, but even localized projects have had significant results.

One of the best examples of an entry-level LeanSigma project occurred at a large midwestern manufacturer of residential entry doors, Pease Industries. On a line that created the resin and glass inserts for high-end residential entry doors, there was a 16 percent defect rate and the line was in danger of being discontinued. First, the consultants knew they had to create flow on the disheveled assembly line. Operators, managers, and consultants formed a team and took waste off the line, eliminated the batch operation, and redeployed operators who were no longer needed on the line. One-piece flow was established, productivity increased by 62 percent, and the defect rate was reduced to 11 percent from an original 16 percent.

It was important for the associates at Pease to experience their work area as a lean environment before we returned with a LeanSigma event. Only then could the associates see the next step. True quality issues are often difficult to uncover in batch mode. In the shuffle of batch manufacturing, it is not imperative to address quality because operators and managers are accustomed to waste and to using inventory to cover up for bad quality. Once a line is lean, workers cannot just accept or overlook problems; they must address abnormality and quality issues. Bad quality in a batch environment is a lot like trying to see the view on a smoggy day. You may be able to see

THE PERFECT ENGINE

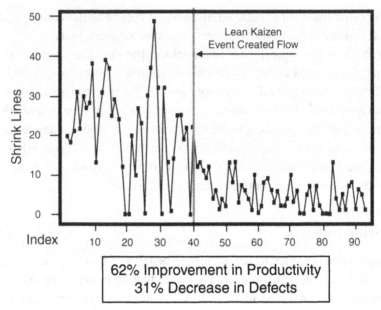

Figure 5-14 Pease teams quickly achieved outstanding results.

a few buildings, but not the entire city and certainly not the horizon. In a lean environment, the air clears and bad product becomes apparent.

Shrink Lines in the Resin

A better environment helps expose the problems, but it does not always uncover the root cause. Here is where higher-level statistical tools helped Pease. After the kaizen week established good flow, about 11 percent of the decorative glass inserts for a wooden entry door still showed a consistent hairline imperfection where the liquid resin was supposed to meet the edge. Engineers at the company called it a shrink line because that's what it looked like, as if the resin had shrunk away from the edge.

The engineers and managers believed that they already knew the answer to this quality issue—humidity and temperature fluctuations in the mold department—and they were trying to decide which environment was best—high or low humidity, cold or heat. They decided, before the

Lean Production System

LeanSigma event, to buy a high-volume air-conditioner to stabilize humidity and temperature. This thirty-thousand-dollar air-conditioning unit was approved and scheduled for delivery within a month.

Prior to the event, the company had collected and recorded temperature and humidity for each unit of product for a full seven months. We refer to such data, collected over a long time, as historical. Properly constructed historical data are very valuable because they allow us to look at trends, to see where processes might have changed over time, and to determine baseline process capability. So it was a good impulse, but there were problems. Pease only collected data when a problem occurred. We did not know what the temperature and humidity looked like for good pieces—only bad. So certain were they that environment was the cause, they limited data collection to those two elements.

Going in, however, the historical data were what we had. So we took a representative sample of the data and created a regression analysis for humidity, temperature, and the combination of humidity and temperature. There was no correlation between those two characteristics and the defect. None. Most of the team that had promoted the need for an expensive air-conditioning unit—and the operators who were looking forward to an air-conditioned workspace—had trouble believing the data, but there they were.

Root Cause

The LeanSigma team, which consisted of engineers, quality managers, an operator, and a complete outsider, went to the shop floor to collect live data. The consultant wanted the whole team to truly see the process and the context of the problem. The team members would see things live that they would never be able to capture as data points. It is one of our core beliefs that you see a lot more from being where the action is rather than being in a conference room, looking at a compiled-data picture of the process.

Our green belts and black belts are taught to collect data in a manner that makes it truly useful. We do not use paralysis analysis. Instead, team members recorded data on percentage of defects by part type, by monthly occurrence, and then looked at "shrink lines" by day of the

THE PERFECT ENGINE

week. That's where we found, strangely, that the defects dramatically rose at the beginning of the week and took a straight drop down after that. Forty-two percent of defects occurred on Monday, 35 percent on Tuesday, 11 percent on Wednesday, and just 7 percent on Thursday (the plant doesn't run on Friday). Then we began constructing a LeanSigma quality map, showing each step of production, including the part where the operator stopped and *smelled* the mold to check for resin buildup.

A quality map shows each step in the process: from raw material to forming, logo application, mold, trim, and inspection. This should be a true picture of the current process, noting all gauging and control methods, tooling and fixturing, any known problems and the design specifications. This becomes a core document, driving the team's focus.

The new information from the enhanced spaghetti map and quality map were then plugged into a cause-and-effect diagram and all possible causes for the defect were listed. Further focused, the team then created a comparative analysis worksheet. This tool looks closely at the what, where, when, and extent of a problem, and then it looks at what the problem *is not*. The comparative analysis is fully supported by data, with teams often performing statistical analysis on data streams to determine what the defect *is* and what it *is not*. The team at Pease could see that "shrink lines"—which they began to call Wilbur to avoid prejudgment—were a problem, for instance. But quantity of resin, swirls, bad resin mixes, and lamination were *not* problems.

The team studied the distinctions between what the problem was and wasn't, when it occurred and when it did not, where it happened and where it did not, and created simple, logical if/then statements. Then they deselected all of the possible root causes. Of those that were left, there were seven different ways of saying one thing: We had dirty molds.

After Analysis, Improvement

Then it was time to test so team members went back to the shop floor where they tested new molds against clean molds and dirty molds. Both new molds and clean molds had no defects and dirty ones had defects occurring at the expected rate. Wilbur was all over the place on the dirty

Lean Production System

	Is	Is Not	Differences
What	Decorative lites	Clear laminates	Exposure to air, less resin, resin, ratio, no mold for lams, 1-pc. glass to 2-pc. glass, clamps, gel time, cure time, rigid vs. plyable, taped borders, sealed at pour
		123's	Treated as gold, very low demand, **frequency of mold baking**
	Shrink lines	Short shot, swirly, bad mix, delamination	Surface vs. mix or fill condition
Where	Demold, swiggle, conveyor, consumer, retail, distributor, Betty	Unit line	Education
Drove focus to mold prep: chi square results showed the day to be statistically significant			Air contact, clamp pressure, improperly cured, **poor cohesion**, contamination, **uneven saturation**, clamp location, blot on edge not in center
			People, **new molds**, cure-time
			Mold design, **styrene buildup**
			Mold sets over weekend w/o ventilation
			Status quo
Molds were cleaned on certain days			Smaller is harder to see, location is harder to see, criteria, defect subjectivity
			Clamping pressure, subjectivity, clamp location, # of clamps, defect criteria

Problem Description: A shrink line is a hairline imperfection or scarred appearance of the molded resin.

Figure 5-15 Comparative analysis.

molds. This was followed by one other curious event that brought home the power of kaizen: As the team was out on the shop floor validating our findings, we watched an operator named Phung Mai separate a mold from glass in a manner completely different from Standard Operations.

Phung Mai's technique, wrong though it was, completely eliminated a hairline defect that had always disappeared after an hour, but was troublesome. Phung Mai's technique was made standard operating procedure on the spot. In theory, we all knew that cleaning and baking the molds would be good. But frequency was not defined and follow-through was not seen as important. Now it is standard work for the operators to clean and condition the molds every ten pours.

By Wednesday, nobody was measuring for a high-volume air-conditioner anymore and that thirty thousand dollars was back in the capital expenditures account. The team members who were reluctant at first to believe that temperature/humidity was not the culprit were convinced by real data that dirty molds were the problem.

THE PERFECT ENGINE

We had collected live data, sifted through it with LeanSigma tools, come to a probable root cause, tested it, and verified the results, all in just three days. Best of all, by current management figures, Pease saved $1,050,000 a year in scrap and manpower reductions by using kaizen and LeanSigma.

MAIC—The Wheel of Fortune

Most quality-focused issues are not resolved quite this easily. Shortcuts in the standard LeanSigma process were taken at Pease when the issue became obvious—a tactic we applaud because we believe in a bias for action. But this kind of detective work, in which we are hunting down a root cause, requires more complex tools.

In general, the phases of such projects follow the MAIC wheel: Measure leads to Analyze, leads to Improve, leads to Control. For a classic LeanSigma quality and variation project, we use a week for the Measure phase, a week for Analyze/Improve (because after the team knows what to do, we never sit around and wait), and a thirty-day followup for Control to ensure that improvements really fixed the problem. There are typically a few weeks in between each weeklong phase, when additional information is collected and tests are run.

The circumstances guide the work, however. Sometimes circumstances dictate longer breaks between phases or shortcuts.

Each phase of an advanced project turns up a list of improvements to be made, uncovering issues that could have remained hidden, causing trouble, for years. And even companies with seemingly straightforward processes discover dilemmas they could not earlier see.

Tasting the Slurry: Cold Spring Granite

Beneath the snowy hills and fields of northern Minnesota, the earth gives up the raw material of Cold Spring Granite's industry. The raw material that has formed over millions of years is perfect, they like to say at Cold Spring, because Mother Nature is their supplier. When stone doesn't come out right, blame the human processing—the quarrying, slabbing, and polishing. In this case, they knew it was the shot saw.

Lean Production System

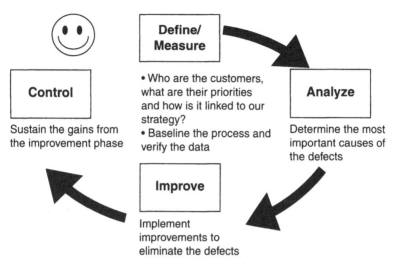

Figure 5-16 LeanSigma project phases.

The shot saw is as big as a house and might contain eighty blades for cutting granite into slabs. Imagine a giant loaf of bread—about five foot by five foot by eight foot—being cut at your local bakery, all at once, into a sliced loaf, and you get the picture. Except, of course, granite does not yield to the knife like flour and butter. So Cold Spring Granite uses techniques perfected over centuries: As the dull blades saw down into the loaf of granite, a gray mud called slurry is fed into the cuts. Slurry is a careful mixture of water, lime, granite dust, and crushed steel shot. That last item is simply steel pellets, like birdshot, crushed up to create sharp edges. Like tiny chisels, the crushed steel shot bites into the granite and makes the cut, guided by the blade.

Cold Spring Granite is probably the world's largest processor of granite. They have been in business for more than one hundred years and a majority of the monuments and mausoleums in your town can probably be traced back to their quarries and slabbing in Minnesota, Texas, and the mountainous gold country of eastern California. Executives knew they could grow their business even larger—outside the known limitations—but not with the quality their Minnesota shot saws were producing.

The dozen huge saws in Minnesota consistently produced uneven results. Sometimes a slab would come off the saw so smooth, it would

THE PERFECT ENGINE

barely take any time on the finishing lines. More often, the face of the stone would be rippled with etched lines, striations that would slow polishing lines to a crawl as workers tried to smooth and polish the stone.

Art Versus Science

Cold Spring first turned to the saws' manufacturer, Italy's Barsanti. In traditions that have been handed down for generations, Italians have special masters who oversee saw operations. It is the master's job to ensure that the saw is operated correctly, that the cut stones are smooth enough to polish, and that the slurry mixture is correct. To test the slurry, the special master tastes the gray mud. Italian saw masters claim they can tell the density, viscosity, and amount of grit—steel shot—just by the tasting.

Cold Spring executives considered creating a saw master position, but they could not imagine including mud tasting in an associate's list of duties or standard work, recalls TBM consultant David Beal, a LeanSigma black belt. "They were thinking of making an offer to an Italian master. But then we said, 'wait. Let's see if we can turn this into a science rather than an art.'"

After further investigation, Beal found that the first issue to address was definitions. In trying to do a Gauge R&R (reproducibility and repeatability), Beal found loose agreements and disagreements, but no metrics. Cold Spring had never defined "rough" stone or "smooth" and opinions could vary wildly. Most operators defined good as "as long as polishing is possible, let them deal with it."

The LeanSigma team first established a zero-to-five-scale definition. Zero was too rough and a five looked very smooth, requiring minimal polishing. If a stone was a three it could pass through the polishing line at 120 millimeters per minute. This was an important metric because 120 millimeters per minute was the designed speed for the polish line. Historically, however, the line was running stone through at a rate of about 80 millimeters per minute. The polish lines were seen as a bottleneck, but the team knew better.

Lean Production System

Quality Mapping

After studying the shot saws in motion and collecting process information—including the fact that each saw runs on an eighty-hour automatic cycle time—the team set about making a quality map. For each step that transformed the raw material, the team noted specifications, control methods, tooling and fixturing, known problems, and any trials or results. The team studied setups for the mammoth saws, how the slurry is delivered, actual slabbing, and then how the stones are removed and rinsed. With each step drawn out in pictures, abnormalities or incongruities leapt to the fore.

In this case, team members found themselves looking again and again at their system for measuring the slurry: a simple gauge, a dipping stick with a one-liter bucket attached, another bucket, and a screen. They found that the operator held out this dipping stick with its one-liter container into a waterfall of rushing gray mud. Sometimes the slurry came in too fast and just leapt back out of the container, leaving it not quite full. But the operator took the container, dumped it into a bucket, and rinsed it,

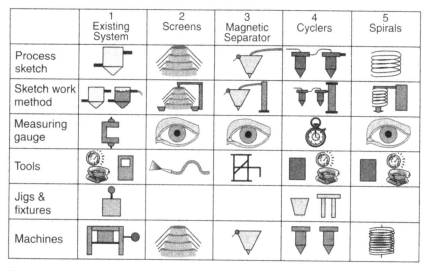

Figure 5-17 For every issue, the team must sketch out multiple possibilities.

THE PERFECT ENGINE

pouring off the excess water until mostly the larger chunks were left. Then he dumped it onto a screen, rinsed again, and weighed the remains. That was supposed to tell the operator how much steel shot, or grit, was in the mix, per liter.

"What we needed was a measurement system analysis," Beal says. "We saw that we had poor specs for viscosity, and poor specs for density or amount of shot. The way that we took the sample, and measured the sample, created more variation than there might have been in the pool of slurry. There was too much variation."

The Veterinarian's Syringe

So the team did a comparative analysis, focusing closely on what, where, when, and how big was the problem. They asked what the defect was and was not, where the defect occurred and where it did not, and when the defect first occurred. And they found something interesting. The defect, the rough striation on the granite surface, did not occur in California. Further investigation found that California was the only site that did not have an automated system to manage the slurry from one cut to the next. Close to both the root cause and the answer, the very modest Minnesota engineers and operators completed their week's report with this conclusion to upper management: "The slurry used in the shot saws sucks." More conservatively phrased, the slurry sometimes had poor viscosity, density, and grit proportions.

"We knew we had to find better ways to measure the slurry," Beal remembers. "We spent a lot of time on this until this one middle-aged engineer, Aleksander—I think he was Yugoslavian—started asking someone to take him to the 'wet.' Finally we understood him to say vet. So we go to a veterinarian's office where he got an enormous one-liter syringe. This thing didn't have a needle on the end; it had something like a three-eighths pipe on the end. Aleksander attached the syringe to a valve in the slurry-feed tubing and got a sample."

With a method of getting good, consistent slurry samples, the team could finally move forward. Each of the three sites—Minnesota, Texas, and California—adopted the same measuring technique. They were looking for 120 grams of small grit per liter. Small grit is the previously ground steel shot that is no longer truly useful to the process. As soon as operators could measure consistently, they found that they were getting about 330 grams of small grit per liter in Minnesota and Texas. California, on the other hand, maintained an average in the target range. On the zero-to-five scale for rough finish, California was also consistently getting fours and fives.

True Grit

Outside California, the ratios of good grit to bad varied widely and were unstable throughout the eighty-hour cutting cycle. The team ran its data through a Best Subset Regression, to find which element—big grit, small grit, viscosity—had the biggest influence on quality; it found that the amount of fine grit was the biggest offender. No question.

The fine grit got in between the steel blade and the stone surface and created striations across the granite face. Also the fine grit filled up the spaces between the tiny chisels of the rough steel shot and the stone, stopping the cutting process. There was too much fine grit, team members agreed. But how do you extract particles from mud? They could add lime, add water, add steel shot. But how would they take the fine dust out? At the end of the Analyze phase, the team set new specifications, sampling methods, and standards to control fine grit, and then prepared to meet again about a month later.

As with all LeanSigma events, the team put creativity before capital as we met to discuss how to remove fine grit from the slurry. The first objective was to come up with seven unique alternatives to removing fine grit from the mud. We finished with thirteen.

First, we drew Cold Spring's current method: the gauge and bucket and rinsing and weighing. They considered magnetic separators to grab all the steel out of the mix. They looked at spirals to spin the slurry and centrifugal force to separate the elements. They talked through screens and cyclones and cyclers and all variations on a centrifuge and then measured our ideas against the relevant criteria. They asked if each solution could meet takt time and budget, whether it was mistake-proof, right-sized, safe, and would cost minimum development time while granting Cold Spring a technological advantage.

THE PERFECT ENGINE

Slurry Dialysis

What the team came up with was a small, mobile unit now called a Slurry Dialysis Machine. Just three or four feet in diameter, it sits on wheels so it can be moved to each saw and plugged in, where it works like a kidney dialysis machine that cleanses a person's blood of impurities. The slurry goes in at the top, where water, lime, and dust are pulled out in the first section. The big grit is separated in the next stage and the small grit all settles to the bottom. Water, lime, dust, and big grit are all sent back into the slurry while the small grit is discarded.

By focusing on the fine grit, Cold Spring was able to increase its quality levels to a consistent three or higher, and the polishing line was up to an easy 120 millimeters per minute. In all, Cold Spring figures the higher quality standards are worth between eight hundred thousand dollars and one million dollars a year to the company.

Even better, says Greg Flint, Cold Spring's head of continuous improvement, supervisors approached him after the MAIC cycle was complete and said they had learned more about their saws and slurry in a few weeks of LeanSigma than they had in the previous fifteen years of running their saws.

Control

Instead of gathering a team for another event, the Control phase is actually contained in the thirty-day homework and it almost always focuses on the most important control there is: standard work. This is the time for the designated team member to rewrite the standard work, or Standard Operations, of every operator affected by the project.

We strongly suggest that supervisors, along with team members, perform thirty-, sixty-, and ninety-day audits of the process to ensure that the improvements are etched in stone. This is also where we return to the teachings of jidoka. Try to mistake-proof the process and then leverage whatever the team has accomplished—all of the improvements—throughout the organization. Because Cold Spring Granite had Californians and Texans on the team, for instance, it was immediately leveraging its improvements across the board. Finally, team leaders must get all financial impact statements approved by the accounting department—ensuring that the metrics can be agreed upon throughout the organization.

A Better Environment

Faster results are only one reason to integrate lean manufacturing with a Six Sigma program. Yes, projects are completed faster and generate better quality with less capital. But there is a human factor, too.

Think about most shop floor and business offices in your town. If the company tries even one improvement initiative, chances are it is trying two or six. In major manufacturing firms, we have seen too many such initiatives fail because they are never really supported by upper management and they clash with other projects and programs. Companies struggle to improve amid a din of voices, caught between competing improvement programs littering the agenda with conflicting priorities. Forward movement grinds to a halt.

Too many manufacturing companies have been caught in the same way between lean and Six Sigma. Not only would the two programs compete for resources, there was also a culture clash. Most lean programs—especially those rooted in kaizen breakthroughs—are centered on teamwork. Too often, Six Sigma had an elitist strain, with black belts left alone to crunch numbers and work on long projects in offices far from the factory floor. With two cultures clashing, little could be accomplished.

What co-author Sharma saw was an opportunity to take the best of Six Sigma, where decisions are driven by data and the more useful tools available in statistical analysis, and integrate with lean. In the end, each strategy informed the other, creating something better than either alone. These are the principles and the synergy that now guide companies as they strive for continuous improvement.

Although lean pioneers like Toyota focused primarily on manufacturing, we have seen successful application of continuous improvement methods contained in the LeanSigma Transformation applied successfully in engineering, order administration, and research and develop-

THE PERFECT ENGINE

ment. There is no end in sight for the application of these powerful methods.

Perfect processes and well-trained, energized, and motivated workers don't just happen. Visionary leaders and sound systems create and solidify change. It's a constant improvement cycle, starting with planning and observation to understand the current state, to organizing and defining the new state, and, finally, implementation. The fun of change lies in implementation, but continued success only comes with revisiting and rerunning kaizen events on last month's or last year's project.

"The new and improved" process and "the new and improved" workforce is never complete. When corporate or market or technology strategies change, it is imperative to have a responsive production process and a well-trained workforce to make the move. We encourage everyone who desires success to embrace change. Because if everything is changing around you and you are stagnant, change becomes a threat lurking over the horizon. However, if change is everywhere and you are nimble, lean, and responsive, you will learn to create enormous value from change.

It is a journey, not a destination.

CHAPTER 6
Lean Ergonomics and Safety

An ounce of prevention is worth millions of dollars of cure.

There is an ugly side of work that all of us have seen and too many of us have experienced: the unhealthy, dirty, boring, and even dangerous places that mark flawed processes and bad equipment. Even today, over 150 years after the launch of the Industrial Revolution, worker health and safety challenges continue to fuel the argument over responsibility for building a better workplace: Is it the government's, the union's, the corporation's, or the worker's job to spot and push for better work conditions?

Further, no industry is immune to the problem—machine shops, steel mills, plastic molders, and raw material processors are traditionally cast as likely offenders, but electronics and light hand-assembly plants, as well as food processing and pharmaceutical sectors, are equally susceptible to practices that hurt workers and fail to build workplaces that enhance the joy of working and enable strong productivity. But for every bad industry image—the scarred limbs and missing fingers among foundry workers, for example—there is an equally exemplary model of work design done the *right* way.

At Honda of America's Anna, Ohio, engine plant, for example, a giant assembly operation and foundry exist side by side in the same plant, virtually odorless and laboratory-bright. In the Pella window and door factories, powerful vacuum systems and ingenious workplace environmental

THE PERFECT ENGINE

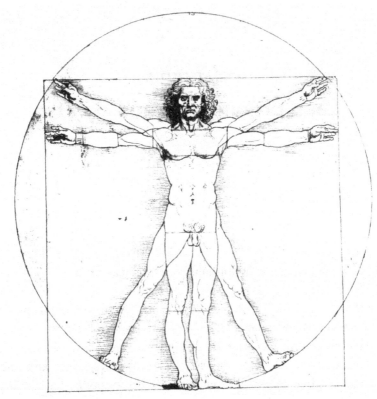

Figure 6-1 *The Proportions of the Human Figure* by Leonardo da Vinci, 1490.

assists eliminate the inevitable problems from airborne sawdust and chemical vapors that must be moved and processed through a variety of operations. And in dozens of consumer appliance assembly cells, workers maintain steady eight- and ten-hour rhythms, performing a variety of tasks without such repetitive injury problems as carpal tunnel syndrome, tendonitis, and aching muscles and limbs.

Ergonomics' Rise

The science of ergonomics—from the Greek word *ergo,* for work—is described by some experts as a black art, a science created out of a shifting mix of classic engineering principles combined with newer and more dynamic human factors. Although commonly accepted guidelines may direct

Lean Ergonomics and Safety

the proper design of a work process—an auto assembly line, for example—practitioners must sift through a database of experiential as well as theoretical and governmental regulatory information to seek the best way to design ergonomically efficient processes, tools, and even products.

Legislated Practice: OSHA and Other Governmental Solutions

"Work-related MSDs (musculoskeletal disorders) currently account for one-third of all occupational injuries and illnesses that are severe enough to result in days away from work," says the U.S. Occupational Safety and Health Administration's (OSHA) chief Charles Jeffress. "It is our top priority."*

Over the years, governmental influence on prevention of workplace injuries and illnesses has grown and waned, as has union influence over this enormous human and financial problem. The proponents of ergonomically good design constituents—union, management, labor, and government—have a common stated objective to reduce or eliminate workplace injuries and illnesses, but their consensus diverges at possible solutions and preventive practices, as well as worker's compensation plans.

Further, no one is clear about the exact cost of compliance or preventive programs, but the reported cost of accidents and unsafe work environments is apparent. According to data from the Bureau of Labor Statistics (BLS), the number of repeated trauma cases in the workplace has declined 24 percent since 1994—to 253,000 in 1998, or 4 percent of all occupational injuries and illnesses in both years. "In light of the . . . data that show MSDs have decreased almost 25 percent over the last five years, one questions the need for bureaucratic intervention," says Ronald E. Bird, chief economist of the Employment Policy Foundation, in Washington, D.C.

But other data show conflicting definitions of the range of injuries classified as MSDs; agreement will not come soon between all parties.

Cost estimates on the extent of the problem range from OSHA compliance cost projections of $4.2 billion to $100 billion (EPF). But, says OSHA's Jeffress, these exponential projections of costs are inaccurate: "These studies predicting high costs are just making bad assumptions,"

---

*Verespej, Michael A., "Election-year Ergonomics, Can OSHA Make Its Proposal Stick?" *Industry Week,* April 17, 2000, p. 51.

THE PERFECT ENGINE

he asserts. "They assume that more workers will be injured, that all jobs will be automated, and that there will be productivity losses from ergonomic changes. We believe that those assumptions are wrong. *Good ergonomics reduces costs and increases productivity.*"

OSHA's proposed ergonomic workplace standards are subject to continued skepticism. Their latest proposal, a six-point program designed to be implemented *after* an injury has been reported, is criticized for not being a preventive program. Under the OSHA Proposed Ergonomics Standard, employers are required to implement a basic ergonomics program that contains these six elements:

1. Management leadership and employee participation
2. Hazard information and reporting
3. Job hazard analysis and control
4. Training
5. MSD management
6. Program evaluation

The LeanSigma Transformation Solution: In the Heartland

We believe these solutions are necessary but they are not forward-thinking changes that will enhance the joy of working. Simple, elegant solutions are needed to make the work effortless and rhythmic.

There is another way. In dozens of breakthrough applications, lean experts point to examples of brilliant, efficient, and far and away cost-effective solutions to everyday ergonomics problems.

At Lantech in Louisville, Kentucky, new stretch-wrap machines are designed with the customer's ergonomic concerns, as well as the assembler's health, in mind.

In Maytag's Cleveland, Tennessee, plant—a legacy site working hard to take the high road with a mix of inherited challenges typical in heavy metal appliance production—the workers themselves have taken the ergonomics challenge to heart and designed smart solutions to ordinary challenges of lifting and moving materials and assemblies. Their solutions are remarkably simple. They make use of basic principles of physics and common sense: Observe the human body's comfort level and design work movement to fit in that range; use gravity rollers and

Lean Ergonomics and Safety

wheels to feed and move materials; eliminate vertical movements against gravity; and use simple machines like scissor lifts and rolling carts to protect delicate human skeletal structures. One of Maytag's new lines is designed with carts to eliminate lifting units from station to station. A former Black & Decker plant in Mexico has made great ergonomic inroads with well-designed cells that replace traditional assembly lines and make work smoother for operators.

In these breakthrough applications work has taken on a different rhythm, a clearly discernible pattern linked to the calculated takt time of customer demand factored by workers' demonstrated best ergonomic capabilities. Where a mismatch of product flow or demand and operator pace is observed, simple changes in product flow or cell design offer inexpensive and quick solutions.

Further, the list of ergonomics improvements throughout LeanSigma installations is as long as workers' creativity and thoughtfulness. All it takes is an ability to see clearly, to study a task or a given operation, and to brainstorm and select the simplest solution.

Lean Ergonomics: Designing Work for the Human Body

The LeanSigma Transformation focuses on a balance of simple flows assisted by selective and appropriate introduction of machines and some technology aids, governed by human issues. The goal, of course, is to make work more enjoyable and less wearing on the human body as we build efficiency. Lean ergonomics smoothes workflow and makes work more consistent.

A common misperception, that changing a traditional work environment to a lean one speeds up work and makes it harder for operators, is quickly dispelled with the introduction of lean ergonomic methods. Not only is the work design concentrated in the normal body, shoulder, arm, and wrist positions, but the tools are modified to enhance normal hand position.

Building Rhythm

Lean principles introduce more variety and different tasks into takt time. Instead of an operator on an auto assembly line, for example, hanging

THE PERFECT ENGINE

door panels for eight to ten hours, building in repetitive-motion stress injuries and boredom, as well as the inevitable possibility of quality problems, LeanSigma introduces a reasonable variety of well-designed tasks into the work cell. The operator uses hands for picking, placing, and moving small objects, as he or she moves around the work cell, thus eliminating the hard work of standing or sitting in a single position for long hours.

Bending to pick up, move, and position heavy loads—the primary cause of lower back strain—is eliminated by simple machines. Reaching beyond the comfort zone for tools and materials is cut out as operators redesign the location and availability mechanisms for frequently used air guns or supplies. Inevitably, visibility to clean line-of-sight manufacturing management is improved, and the more workers clear out extraneous and erratic operations from the flow, the more the entire process becomes an observable, rhythmic sequence.

Work distributed over different parts of the body creates a pleasant rhythm that eliminates the traditional work pattern of "wait, work, wait" played out in short, jerky motions broken by minutes of boredom or panic as line pace overrules the human element. We have learned from the study of human exercise and aerobics that moving in rhythmic, consistent ways always beats erratic, strenuous attempts at performing work. An ideal takt time of about sixty to six hundred seconds offers the optimum opportunity to include multiple tasks connected in a smooth, consistent rhythm. This approach also keeps the standard work routines simple and easy to learn, which makes adherence to work practice easier.

Designing Product for Good Customer Ergonomics

Good ergonomic design is not confined to improvement of production processes. The way product is packaged, transported, uncrated, installed, and even used can have equally serious ergonomic impact. Breakthrough projects at Lantech have demonstrated the parallel importance of product and process design for good ergonomics all the way through packaging and trucking to unloading and installation at the customer site.

Quality defects, repair rates, rapid installation to perfect run times, and

Lean Ergonomics and Safety

other issues previously classified as unknown quality problems are traceable to human assembly, handling, crating, and uncrating, particularly in complex machine product offerings. Clearly, every single operation, from raw materials through setup and initial trial runs, must be designed with the best human factors in lean ergonomics in mind. The way a machine is designed for assembly, for example, with all the issues of movement, positioning, and efficiency, is just as important at the other end of the process, where packing, wrapping, storage, and eventual installation and use affect how well the product operates in the eyes of the consumer.

Economic Benefits of Lean Ergonomics

There's another good reason why manufacturers must design product and processes based on lean ergonomic principles—the financials. Building good products and processes that smooth flows and eliminate human wear and tear and injuries makes good business sense.

In 1993, early on its lean journey, Lantech learned something new about the benefit of being lean. Data showed that in the first six months of the previous year, from approximately January to July 1992, the company had spent about seventy-three thousand dollars for medical claims—injuries to hands, backs, feet, and other parts of the body. The paperwork to track, complete, and process these accidents was so substantial that one full-time employee was dedicated to medical claims.

But Lantech experienced the welcome benefits of kaizen applied to human factors early in its kaizen journey. During the first six months of its journey, the cost of claims dropped from seventy-three thousand dollars to eleven thousand dollars, and the claims processor started to complain about not having enough work to do.

Co-author Sharma contrasts this six-month turnaround to another company's experience. Sundstrand Aerospace was considering the launch of several kaizen projects, but its union proved to be a barrier to getting started. Union leaders were nervous about cells, and they questioned the safety of workers performing ordinary functions in cells. Their questions demonstrated a lack of understanding of the principles behind good work design; nevertheless, their concerns had to be dealt with be-

153

THE PERFECT ENGINE

fore kaizen could continue. The best method to overcome this objection was to tour the current workplace and point out the body stress that was being created by supposedly safe work practices and then to demonstrate the ergonomic improvements that would be made as part of the lean implementation.

A strong visual statement is often the best way to approach misperceptions and fears about changing workers' tasks and their work station design. One particular video taken in a Japanese automotive sound system speaker assembly cell illustrates this approach. In this cell a woman worker works to a twenty-second takt time, performing four or five different operations as she moves in a circular pattern through the steps; amazingly, the process is so well choreographed that between 8:00 A.M. and noon, four hours later, the rhythm of her work is relatively unchanged—only a one-second difference separates the morning cycle from the noon pattern. As one watches this video, it confirms beyond any doubt that if the body moves in stress-free rhythmic patterns, we see a dance routine performed at work.

Long-Term Costs

Dr. William Mallon, a former PGA pro golfer turned orthopedic surgeon now based in Durham, North Carolina, witnesses the enormous human and financial costs of bad ergonomics every day. In his practice, Mallon sees a mix of upper extremities' (shoulders and hands) overuse injuries and worker's compensation claims, as well as the typical severe emergency room–type injuries.

Of his seven or eight hundred new cases per year classified as upper-extremity injuries, 30 to 40 percent, or two hundred to three hundred per year, represent worker's compensation cases. The average age of the workers is forty, and claims are evenly divided between men and women, although typical problems for women include carpal tunnel and forearm and wrist tendonitis, both of which can take a "long, long time to cure," says Mallon, "and are very difficult to treat and reach a positive conclusion. Workers generally arrive in pain, and they cannot work." For men, typical problems would be more severe, emergency-room cuts, lacerations, and chronic injuries like rotator cuff tears.

Lean Ergonomics and Safety

High Costs, High Cure

Of Mallon's two hundred to three hundred annual new cases, the surgeon estimates that *fully 30 to 40 percent are preventable,* for both men and women. Although the specific injury may have occurred at work, doctors cannot always implicate the workplace as the *sole* cause. In fact, most men in their forties or fifties have at some time in their lives injured a shoulder rotator cuff, but under a heavy load in a typical work situation, the cartilage finally tears, sending the worker off for expensive repair and rehab work.

And the costs are never a one-time expense, because major injuries like rotator cuff repair require surgery, physical therapy, rest, and continued strengthening of supporting musculature. Cost to repair the typical workplace injuries ranges from three to more than twenty thousand dollars of high-tech surgery. Carpal tunnel average repair cost, per hand, is three thousand dollars; rotator cuff is six to seven thousand dollars; a broken wrist is five to six thousand; back injuries, the dreaded lingering problem so frequently experienced by men who lift weight, can easily reach the twenty-thousand-dollar mark for surgical repair of several ruptured disks requiring fusion, not including rehabilitation, therapy, lost wages, and disability pay.

Adding in the postoperative costs to a rotator cuff tear, notes Mallon, raises the costs alarmingly higher. A typical rotator cuff tear repair requires two to four months of rehabilitation, plus the cost of therapy—add another few thousand. Soon the total hits ten thousand dollars for medical care, at minimum. Add the cost of making the diagnosis with MRI and other x-rays for one to two thousand, and the bill easily reaches twelve thousand dollars.

Depending on the severity of the injury and the age of the worker, lost work time can last from six weeks to six months, at a cost of anywhere from five to twenty-five thousand dollars of lost production. Finally, in Mallon's state of North Carolina, the injured worker receives a lump sum settlement based on rating the disability's permanent impact on the worker's life: A permanent partial impairment rating, PPI, estimate is based on what percentage of normal motion the worker is expected to regain.

"It gets worse, much worse," says Mallon. "If the injury happens to an average male high school graduate who works on a tire assembly line, for example. That worker will probably never go back to lifting heavy loads, especially on a job designed for a twenty-year-old. The bottom line: a worker who cannot work and stays at home, collecting salary until age sixty-five. I estimate that this worst-case scenario happens 3 to 5 percent of the time with industrial injuries. The worker and his family, as well as the employer, are stuck if he cannot be trained for another job."

As we begin the twenty-first century, work injuries are a problem for all businesses, not just heavy manufacturing. Health insurance costs rise annually, almost always outpacing inflation, and worker's compensation insurance continues to represent a significant cost to large and small businesses alike.

In the first half of the twentieth century, worker's compensation injuries typically involved acute injuries sustained by a single traumatic episode on the job. But in today's work environment we now see the addition of a relatively new challenge, repetitive stress injuries, complex problems that we want to eliminate, rather than repair. Repetitive stress problems are fast becoming a major source of worker's compensation disability payments.

"Don't Operate on the Foot, Operate on the Shoe"

Prevention, rather than cure, should be the goal of all enlightened cost-effective producers. But how can LeanSigma leaders incorporate simple and effective ergonomic principles in their strategy for workplace redesign?

What Are the Principles of Lean Ergonomics?

A few simple rules guide the best installations, beginning with the surgeon's advice: "Don't operate on the foot, operate on the shoe." Simply make the workplace fit the associate—don't operate on the patient, but fix the workplace.

For example, in any assembly line men and women of different height and body type work side by side. Of two people performing the same

Lean Ergonomics and Safety

job, one, a six-foot-five-inch man, has greater range of motion and more strength than a five-foot-two-inch woman working next to him. Yet the line is designed for a one-size-fits-all worker.

The lean solution is adjustable work stations, platforms that move to various worker heights. The objective is to perform work in the normal "strike zone," from the middle of the thighs to an area below the shoulder—because bending over to reach lower objects, or repeatedly reaching above the shoulder, will eventually cause injuries and strains. It's a simple principle, but one that presents manufacturing flow experts real challenges.

Guidelines for Lean Ergonomics

Lean ergonomics refers to the concept that if one eliminates factors in the work environment that may cause injuries, ergonomic costs drop as productivity rises.

What makes the study of ergonomics so challenging and difficult to define is its broad reach into engineering of tools and machines at work, and the way they interact with the human machine in a variety of situations. The human machine is very complex and varies from employee to employee, making absolute analysis of all situations nearly impossible. For industry, understanding and reacting in a positive, systematic way to relatively new issues such as repetitive stress syndrome offers tremendous potential for improvement.

Seven simple guiding principles apply to all businesses and manufacturers that want to think about ergonomics' role and to design and modify their work systems to minimize the chance of work injuries:

1. Emphasize acute safety principles at all times.
2. Whenever possible, fit the employee to the job.
3. Fit the workplace to the employee rather than forcing the employee to adapt to the workplace.
4. Design the work environment so that neutral body positions are maintained most of the time.

THE PERFECT ENGINE

5. Redesign tool handles, like the "Beer Grips," to promote normal hand and wrist positions that reduce stress and injury.
6. Vary the tasks performed, including job rotation, every two to four hours.
7. Cater to the human body; make the machine servant to the human.

---

1. Safety Principles

Although repetitive stress injuries are now endemic, they are not completely understood, and acute injuries—loss of fingers and severe lacerations, broken bones, eye injuries, even fatalities—still occur. The workplace needs to be designed with the possibility of all these accidents in mind. Workers should not, for example, work at heights at risk of falling, without adequate safety apparatuses for support—harnesses, railings, or other barriers—in place.

High-speed or high-impact machines—automotive stamping presses, for example—should always be designed to not operate unless hands and upper extremities are removed from the work area. And of course, protective eye, ear, head, and foot gear needs to be available and always in use.

2. Fit the Employee to the Job

Recognize that not all employees are suited to all jobs; personnel as well as manufacturing management needs to understand and adapt to this key ergonomics guideline. Further, it may be necessary to consider legal implications before fully executing this principle. It is important to remember, therefore, that certain jobs require strength, height, or manual dexterity; if an employee is not physically suited to the job, he or she should be redeployed to another position where his or her talents and abilities will be better used.

3. Fit the Workplace to the Employee

The height of the average American female in 2000 is five feet four inches, and the average American male is five feet ten inches; clearly, for workplace design, height and positioning of tables, platforms, a "one-

Lean Ergonomics and Safety

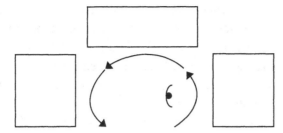

Traditional workplace design required workers to go to the machines.

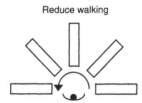

In a lean layout, unnecessary walking is eliminated and parts loading and unloading is brought close to the operator.

Figure 6-2 Machine load and unload.

size-fits-all" approach won't work. Adjustable height platforms, hydraulics, and certain ergonomic tools allow for adjustments of tool grip to varying hand sizes and strengths. Certain tilting and scissor lift tables can improve the ergonomics and adjustability.

Work stations should be designed to fit the right height, depth (no more than eighteen inches), and size of table. Limit use of horizontal surfaces, because they tend to allow work to "spread"—basically look to provide simply a machine on legs or wheels.

For large and complex machines, put the guts of the machines into the background, while machine loading and unloading can be performed close to the worker, in an ergonomically friendly manner.

4. Neutral Body Positions

Most body joints have a wide range of motion, but the extremes of these motions are not usually comfortable, and when the joints are repetitively stressed in extended positions, chronic injuries result.

THE PERFECT ENGINE

Design work and workplaces to maintain neutral body positions. The back should be relatively straight, with a slight flex at the hip, rather than bent over for long periods of time. The arms should hang at the side, rather than being elevated overhead or raised repetitively to shoulder height. And the wrists should be placed in neutral alignment, rather than bent or pressed at unusual angles.

Use autonomation—automation with human touch—for dirty, dangerous, and tedious jobs. Design work for autonomation, but be sure the worker remains in charge. Use small, simple machines in place of robots and "monuments."

The Citizen Watch Company in Japan had at one time 120 people on a U-shaped line assembling watches that traveled over a moving conveyor. Although management wanted to reduce the cost of this line, to automate the entire process would have cost $2 million. The alternative, autonomation—using a small pick-and-place robot instead of full automation—brought the number of operators down to five, at a total cost of only forty thousand dollars. The five operators represented the cost of doing autonomation; they responded to problems, set and maintained a pace, and understood the flows. Full automation would have been an expensive solution, including complete on-site engineering support. Further, hard automation would have required complete reinstallation of tooling to support model changes, which happen rather frequently in that industry.

Design the work to be performed in the "ergonomically friendly zone"—no hands above the shoulders or below the belt, and not more than 10 percent beyond the body line. Reinforcing mechanisms include tables that are as small as possible, positioned at the perfect human work height.

Design work stations in size and placement so that all tools and materials are within easy reach, gravity-fed if necessary, with no extended bending or stretching to locate tools or materials. Identify and position all required tools within easy reach—pencils, screwdrivers, fasteners, and so forth. Use shadow boards to maintain correct placement. Be aware of pacing as well as parts presentation, and as work progresses, ensure that parts are presented where they are needed.

Container sizing: Provide the minimum amount of work and mate-

Lean Ergonomics and Safety

Figure 6-3 The worker is in a comfortable neutral body position with a slight bend in the hips and the box held close to his body.

rial to the cell by limiting container size to a day or hour's worth of work. The right container size sets the shop rhythm. Rather than bins, use chutes made of cut PVC pipes that can be layered up to conserve space.

5. Redesign Tool Handles

For an operation requiring much force, provide leverage-enhancing hand tools or a press, or a lever with a handle. At Critikon in Connecticut, when associates manufactured a blood collection vessel, they found that attaching a rubber catheter to the vessel was tedious. Their fingers went numb, and the operation was still difficult to complete. The kaizen team designed a simple lean ergonomic device, a wooden device at the bottom of the tube, to act as a lever and make it easier to attach the rubber.

Tool application: Design application of tools for downward motion. Using a screwdriver, for example, the worker should stand and screw *down* on the tool, rather than lift upward and twist. Hang tools at a con-

THE PERFECT ENGINE

Figure 6-4 An illustration of smart materials presentation.

venient height, rather than positioning them in a drawer, and mark them with color codes for easy retrieval.

Rethink "standard" tool design. Never accept a standard tool as the only choice provided by the manufacturer, particularly if there is more than one operation involved. If the operator must tighten a nut and a screw, as well as a hex head, rather than trying to locate a wrench, an Allen key, and a screwdriver, design a multipurpose three-in-one tool to adjust all these and eliminate extra steps.

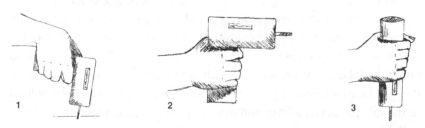

Figure 6-5 (1) Incorrect hand positioning. (2) A small change in hand positioning saves the worker from injury. (3) Redesigned tool keeps wrist and hand in right position.

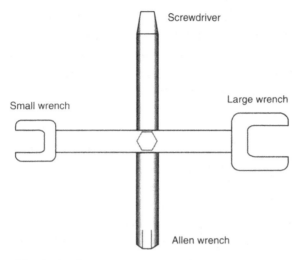

Figure 6-6 Combination tool.

6. Vary the Tasks

LeanSigma Transformation improves business profits by maximizing productivity, but it also reduces the incidence of repetitive stress injuries by designing work simply to be completed the right way. By allowing the employee to participate more fully in the manufacturing process, and by varying the work, kaizen is a very powerful and effective tool to decrease the incidence of repetitive stress injuries. When we combine tasks to the takt time (approximately sixty seconds or more), more parts of the body are engaged in work without rapid repetition. By adding job rotation every two to four hours, companies can avoid any possible stress on certain parts of the body, and reduce boredom as well.

7. Cater to the Human Body, and Make the Machine Servant to the Human

Lean ergonomics, using kaizen to rethink the way we design work, is a critical element of every manufacturer's portfolio, because protecting the health and well-being of workers is as important as producing strong revenue streams and higher profits. Bad ergonomics inevitably result in worker stress, and worse. Productivity improvements are limited when companies allow poor workflow practices to continue. Simple machines, devices like scissor lifts, gravity feeds, tools positioned for quick application, rolling carts, and ramps, form the basis of sound lean ergonomic work design. The basic ergonomics ideas apply across all industries—from metal cutting to automotive assembly and electronics.

First and foremost, we design human-friendly work for the person, to make work enjoyable, and stress-free. Second, we want to create more consistency, repeatability, and reliability in the system. And third, we know that lean ergonomics makes sound financial sense because ultimately, when a society ignores these issues, they cost even more in muda—wasted materials, wasted times, down time, and medical and compensation costs. The objective is to build a system that consistently builds product in a rhythmic way that will not exhaust the human body after eight hours, and that eliminates work injuries.

Lean Ergonomics at Maytag

Maytag operates in a world of heavy metal processing, assembly operations, and material movement that can present an ergonomic challenge for even the most creative work designers. But Cindy Slater, production unit manager of Plant Two of Maytag, Cleveland, Cooking Products operation, responsible for cook tops and wall ovens, thinks kaizen has boosted her area's response to this challenge. Four examples illustrate.

Creaform Cook Top Carriers

Before Maytag implemented lean ergonomics, cook top units were presented in large and bulky steel carriers weighing twenty-five to thirty pounds per piece. The cook top unit sat on the carrier and rode around the assembly line. The heavy metal carriers were, Cindy recalls, "not very operator-friendly—we had operators on the inside and outside of the line, and they had to bend down to do wiring and other operations on a unit. Several of the operators had tasks that included sitting and standing, so they were up and down all day."

Lean Ergonomics and Safety

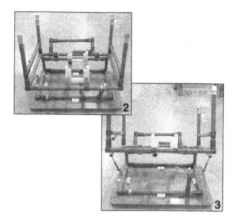

Figure 6-7 The original carrier (photo 1) was difficult for the operator to use. But in photos 2 and 3, the operators rebuilt the carriers with lighter material and hydraulics, making the carriers easier to use.

Two associates on the line designed a lighter creaform carrier (PVC with metal on the inside frame), supported by hydraulic shocks and mounted on a swivel lazy Susan platform. Now the operator can turn her unit 360 degrees around, allowing the operator to perform tasks on the backside of the unit. Further, with the hydraulic shocks, the unit can be lifted or lowered to the height of the operator, with no bending or stooping. The bottom line? Maytag workers realized healthier and faster performance of this operation, at a cost of seventy-five dollars for each new carrier. Time savings realized have allowed operators to tackle more than one task.

Coffins on the H Line

Cleveland's H line produces a forty-eight-inch triple cook top with three cooking bays that is built directly into its wooden shipping crate. The first step in the assembly operation was for the operator to pick up a fifty-pound wooden crate and position it on the line for assembly of the Prostyle unit. The size and weight of the "coffin," however, required two operators to lift and position seventy units in every ten-hour day, or one unit every eight minutes. Clearly, this job represented muda, as well as stressful use of the body.

THE PERFECT ENGINE

Figure 6-8 The original workplace design (photos 1, 2, and 3) was stressful on the operator's body. In photo 4, the new design keeps the operator from doing heavy lifting.

Operators designed an ergonomic solution that improved productivity. They built rolling carts on which fork truck drivers place the coffins, which are then rolled over ball rollers to the line. Crates are delivered to the supermarket from the supplier, one coffin per cart. The eight creaform carts cost a total of four hundred dollars.

Finished Units

Operators at the end of the cook top line, where previously two operators lifted and stacked completed units in stacks of eight, redesigned their job. Now a vacuum hoist performs the heavy lifting. This ergonomic innovation reduced the need for two operators to pick up and stack a fifty- to seventy-five-pound unit over four hundred times a day. The hoist also eliminated one person from the packout area.

Part Presentation and Unpacking

All day long, a waterspider (a worker with responsibility for replenishing parts on a lean line) keeps the motor station (and other work stations on a line) supplied with ten different parts. He reaches into boxes, lifting and positioning the required parts in a flow rack. Each of the four cook top lines hosts one waterspider who keeps the mix of purchased and fabricated parts moving, but as he reaches farther down into the crate to pick

Lean Ergonomics and Safety

Figure 6-9 Parts were kept in large supply racks—an arrangement that required the waterspider to reach into boxes and to lift and bend below waist level (photo 1). In photo 2, the waterspider's replenishment route is changed, freeing up space and saving time.

parts, the waterspider finds himself bending below waist level and straining to retrieve parts.

Operators studied the flow and the waterspiders' replenishment route and came up with a solution that freed one waterspider. One person with a rolling cart, costing about $125, moves one hour's worth of parts per line to fill line A, replenish the cart, fill line B, and so forth. Moving to the rolling cart method freed seventy-seven square feet of floor space on each of the four assembly lines, as motor parts were consolidated from twenty boxes to nine. All parts are now supplied in correct sequence to the lines, and waterspiders are no longer required to bend and reach for parts unloading and stacking.

Maytag also has added fatigue mats at every work station for the operator to stand on instead of standing on concrete for ten hours a day.

Job Rotation

After lean manufacturing has been implemented in an area for six months, production workers go through a certification process. In order for an area to be certified, it must meet certain requirements, one of which is rotation. Each operator is required to know and be able to do at least three operations (the operation before and the operation after his or her own operation) on the line. Rotation has decreased repetitive motion injuries, fatigue, and boredom. The rotation is done twice a day, so that each day each operator performs at least two different operations.

THE PERFECT ENGINE

Lantech's Lean Ergonomics Journey

Lantech, a three-hundred-employee international, privately held stretch wrap manufacturer based in Louisville, Kentucky, is a technology-driven company with over 120 product-design patents and a true belief in and understanding of the power of kaizen. In over eight years of working with lean principles, the company has experienced dramatic productivity gains—15 percent per year for seven years straight—and significant sales growth.

Lantech's current strategy is to expand the product line into secondary packaging lines, not just stretch. But along the way, Lantech CEO Pat Lancaster has made some incredible discoveries about why customers are eager for their product.

"Since five or ten years ago," says Lancaster, "whether the realities changed, or just the popular accounting exercise for capital purchases changed—we have noticed that our ability to sell a piece of equipment has shifted. Customers used to make a purchase decision based on whether a piece of equipment lowered material or direct labor costs because of its direct functionality."

"But today," continues the CEO, "people are buying more equipment with the overall fit to the manufacturing flow—companies are buying machines to make their processes flow and to solve ergonomic issues." For an innovative young organization that prides itself on producing its own product in an ergonomically sound way, this is a great market shift.

Parts Presentation Challenges

Some Lantech smart solutions arise and take root on the shop floor. A recent one involving parts presentation cards, a brainchild of a Lantech cell team leader, David Crain, eliminated much bending and stretching to pick up parts. Crain realized that in a cell using over one hundred small parts—fasteners, screws, bolts, and washers—stretching to retrieve the right part was an awkward job in itself, one that might be done faster with less muscle strain.

Crain's solution, so obvious it belied conventional thought, was to arrange the parts on a tent, each piece at the right level, always within

Lean Ergonomics and Safety

reach—no more bending and stretching, even for heavier pieces like motors, pulleys, and drive belts. Initially, Crain's idea was installed in a feeder subassembly cell making major components for the S straddle machine, but later incarnations found "the tent" in other applications.

Ergonomic Assembly

Another, larger ergonomic opportunity presented itself at Lantech. The company produces large electrical control panels, some as big as six feet tall and five feet wide, stuffed full of components like plcs (programmable logic controllers), circuit breakers, and motor starters. The box looks complicated, and assembling hundreds of wires and screws the old-fashioned way wears on the operator because he or she must work at all levels, on all sides of the panel, bending over, working horizontally, and reaching behind or climbing a stepladder for out-of-the-way connections.

Lantech runs dozens of kaizen events every month, and the panel solution was found during one of them. For less than two hundred dollars, kaizen team members designed an ingenious vacuum plate mount attached to a hydraulic cylinder that moves the assembly to the right level. With a simple foot pedal, the operator can raise and lower the two-hundred-pound panel.

Ron Hicks, VP of Operations at Lantech, loves this solution: "We put the operator on a raised platform. When he hits the foot pedal, the operator can be easily working on the top of the panel, and when he needs to work on the middle, to put the panel at the optimal level, he hits the pedal again. We've eliminated reaching over the head, working on hands and knees, standing on stepladders, and lying flat to assemble flat surfaces, which is what you will see at every other panel shop everywhere—all ergonomic nightmares. In fact, it becomes subconscious—the operator can constantly and quickly adjust panel height up and down."

Hicks understands that not only do these two kaizen solutions eliminate operator wear and tear, but they remove muda as well—wasted time, as well as wasted motion expended by operators as they bend and stretch to compensate for bad parts presentation. "We have a million other examples," says Hicks, "including ingenious devices using scissor lifts, power tools, and rolling racks."

Part of the work of creating sound LeanSigma production plants or

THE PERFECT ENGINE

lines is preparation. The best process will be designed from consideration of technical issues—how the components fit, what equipment and processing steps are required—and bill of material structures, as well as the human element. In the LeanSigma Transformation philosophy, people design themselves into the efficient process, and people—with their human limitations and capabilities—need ergonomic factors to be considered in advance of every new manufacturing project.

Safety and Ergonomics Take Center Stage

Lean ergonomics will continue to offer smart companies the opportunity to rethink the way they build, package, and ship products well into the next century, because everything that workers touch needs to be designed for human capabilities and limits, as well as process efficiency.

In some plants, new employees spend the first day of their "welcome aboard" training learning about safety equipment, at the same time they learn about work processes and Six Sigma quality approaches. Safety for some organizations has become almost routine.

Most legacy processes were designed with little thought to better productivity and the elimination of boredom and repetitive injuries. And in fact, every day, companies that have adopted the Lean Production System are discovering new ways to make work go smoother and easier, to eliminate the possibility of long-term health problems associated with badly designed processes and work methods.

It's Not Just in the Factories

Workers in office and high-tech environments are also susceptible to catastrophic injuries as well as problems that are harder to define, problems that emerge slowly over a number of years. Sometimes the causes are undetected and subtle, not even noticed by the worker himself.

Less Visible Injuries and Health Problems

And what about the hidden injuries and health problems that accumulate over the lifetime of the average production worker or white-collar associate? Problems like work-related psychological stress, problems caused by intensive cell phone or computer terminal and keyboard use clearly affect many more workers today, but the root causes and possible solutions to these more difficult problems are difficult and costly challenges for everyone.

Visual Ergonomics

The human eye, for example, was not designed for uninterrupted eight-hour stretches of focus from one foot away on a colored graphics terminal, and yet every day millions of workers sit for entire shifts at terminals or keyboards, sometimes outfitted with headphones, and often isolated from others by cubicles and dividers. These working conditions are not the ones our bodies were designed for, and one wonders about their bigger impact on workers. The human body, a product of millions of years of evolution from hunter-gatherers, and a few short years of genetic modeling, was probably not designed for a constant assault of electronic signals and communications.

Multitasking and Ergonomics

While multitasking may come more naturally to some workers, transaction speed in human brains has its natural limits. Interestingly, research on the difference between male and female brain structures seems to point to significant differences in basic levels of communications response capabilities. Women workers may have a higher bandwidth, and a better ability to filter and prioritize simultaneous data input as well as extraneous information. Some researchers have concluded that women are naturally hypercommunicative, multitasking experts, while men seem to be more focused on a narrow list of tasks. In fact, these genetic differences may be more apparent, almost exaggerated, as we move into more hyperprocess-based work and integrated communications systems.

It is possible that just as physical size and strength differences created differentiation in labor classification systems and pay scales in old manufacturing plants, there will be some further distinctions made in white-collar work. In the nineteenth-century brick mill, for example, freight handlers were expected to heft different loads than weavers, and pay scales and work hours were different; human capabilities in communications and multitasking should be considered by new manufacturing and service process designers. Where workers are expected to do more, they should be compensated, and where stress and boredom begin to take their toll, processes need to be redesigned to be more human-friendly.

New Process Ergonomic Opportunities

Managers in the new manufacturing plants and newer services businesses need to understand and improve their production and white-collar processes, just as plant managers and supervisors in brick-and-mortar plants worked to keep the machines running without losing workers. It's simply a shift in emphasis, but the basic investment in human health and well-being remains equally important among enlightened corporations, no matter what they produce.

Ergonomically sound processes make sense to lean practitioners. Not only do they prevent costly medical problems and long-standing injuries, but they move people through their work faster, with less excess movement, and they more appropriately match a worker's pace to the true consumer demand. With selective and appropriate introduction of simple technology assists, the powerful combination of human skills and basic machines—wheels, rollers, levers—is continuing to take lean leaders like Maytag, Lantech, their customers and supplier partners, and others into a more productive, enlightened era of manufacturing excellence.

CHAPTER 7
Design for LeanSigma

The Snakelight Experience: Notes from a Team Member's Journal

It's barely morning and the team of engineers, production operators, and would-be cell designers shake themselves into a half-awake state. For some things, black coffee just isn't enough and this is going to be one of those days—or weeks.

Our group has been selected to learn lean design by doing. We are assigned to three- to five-person teams. My group, the Red Team, has only three members. The others are both engineers, familiar with lean manufacturing and eager to learn this process.

We've heard this process unleashes previously hidden creativity and for that reason alone, I am ready to try it out. The Red Team's assignment is to design a new production process using current product design and component pieces for a high-volume consumer product, Black & Decker's Snakelight. It's a three-foot slithery device whose main body is an articulated spine leading to "eyes"—a battery-powered lens and reflector assembly—packaged in molded plastic for easy impulse buys. Although we are instructed to make note of any design opportunities we spot, the immediate task at hand is to design a process using current parts—no changes in the Bill of Material allowed.

THE PERFECT ENGINE

Step one, disassembly (we are learning to work backward), reveals an astonishing number of molded plastic and bent metal pieces, some of which, particularly the articulated spine, we find most uncooperative in our efforts to pry apart. Next we build a fishbone diagram from the disassembled parts, showing the points of assembly. Each of these we label.

Now comes the creative part. Despite the fact that each of these assemblies probably was put together using a standard set of operations determined by trained engineers, we are tasked with the fun job of dreaming up seven different ways of assembling the pieces. Putting the reflector on the head of the light, snapping the clear lens cap on the assembly, stringing together the vertebrae of the plastic spine—all must have seven ways to come together. We sketch each of our seven dream assembly operations—no words allowed—and we specify the type of machine used to perform the operation—hands, feet, even the mouth—and the gauging method, including jigs and tooling.

Some of our ideas are wild, Rube Goldberg expressions that spell mechanical disaster, like the complex fixturing device envisioned for final packing and shipping. The crazier options inevitably arouse interest, but ranking them by various criteria—amount of capital required, operator involvement, poka yoke (mistake-proofing), and jidoka (automation with human touch)—reduces our seven ideas to three workable solutions.

We learn to deselect options because they lack flexibility, or because they would be expensive and time-consuming to change. The goal is not to design the perfect, high-tech all-purpose assembly process. We're going to design a quick and simple device instead, a design that will grow with us and work with human hands.

We work through the fishbone diagram and envision how product will assemble to takt time. It forces us to work to operator rhythms, not a machine's. Along the way, our checklist reminds us to specify simple tools and materials that we will use to build the cell—glue guns, staples, cardboard, Styrofoam, and tape.

By the end of Day One of hands-on work, the Red Team seems to be

Design for LeanSigma

locked in creating elegant and perfectly executed drawings of seven, then three solutions. The Blue Team, however, an assemblage of former operations managers and one-time expeditors, is impatient—critical evaluations and rankings quickly give way to the group trip to Home Depot. Loaded down with credit cards and a long punch list, the team neglects to notice a stash of tools, foam board, and other design goodies piled in the laboratory corner, but never mind.

Next morning, the sun rises a bit more slowly as the two teams are faced with the frightening immediacy of the task ahead—build a simulated production line that meets the objectives of takt time and ergonomics, and follows the path chosen yesterday after the creativity sessions. The Red Team's former tool and die maker, Dave, the grandson of a North Carolina blacksmith, assisted by John, a former Black & Decker engineer, move slowly but surely to lay out the process flow on worksheets.

Dave and John make no move until their flow has been elegantly and thoroughly drawn. In effect, they are creating the standards for this cell from their drawings—true believers in the idea that "90 percent of the work is in the preparation."

The Blue Team, however, having tripled Home Depot's daily revenue goals by assembling an impressive pile of tools and materials, is agitated. They ask themselves first quietly, then loudly and clearly in a room where tension is building like scrap metal flying off a grinder, "What the hell is wrong with the Red Team? We're on a deadline; we're never going to make the Friday launch date! Put down the blasted pencils and pick up a saw—let's get this line built, let's get these parts moving, let's d-o-o-o something!"

Dave and John, imperturbable, remain bent over their process drawings; unmoving, they quietly assemble detailed diagrams for the new Snakelight flows. They are convinced that planning for the takt time goal is important, and that the time for throwing up a simulated cell will come—eventually.

Dave gives us his country wisdom observation, handed down from his grandfather, the blacksmith: "Let the parts speak—the parts will tell you how they want to be handled and assembled." It

THE PERFECT ENGINE

comes in handy when we attempt to feed wire in from the wrong end. And when we struggle with lens reflector alignment, gravity and the parts themselves show the best way.

Sooner rather than later, however, tension and frustration spark small bursts of activity—a foam jig here, and a cardboard tube feeding lens assemblies there. Best of all, operating in mime—performing the actual assembly motions in the lean mock cell, is a real eye-opener for everyone who steps up to the first station. Entrenched habits—working in batch, rather than completing a single unit, bad parts presentation, and even moving too fast—quickly reveal themselves because for every team member, this is unfamiliar territory, truly a time to reinvent the process.

What Is Design for LeanSigma?

A Goldmine

There is a great, unexplored opportunity on most production floors, a veritable goldmine. Ninety percent of all process efficiency—speed, productivity, amount of waste, and flexibility—is locked in even before the first prototype has run across the line, and well before tooling has been ordered or equipment and materials have been pulled together for first-volume runs. These elements represent billions of dollars in sales revenues or production costs. So why is it that most companies have yet to take advantage of this incredible savings opportunity?

Traditional manufacturing practice makes products and workers fit the production process that is already in place. Manufacturing lines are not designed and tested in advance of full production to meet the principles of lean. In the automotive industry, for example, lines are retooled for new models, but most companies have not discovered the powerful production preparation and product design tool that is Design for LeanSigma. They continue to design complex products and inflexible process steps into their production processes. Their continuous improvement initiatives, like kaizen and quality circles, are therefore always limited to after-the-fact improvements.

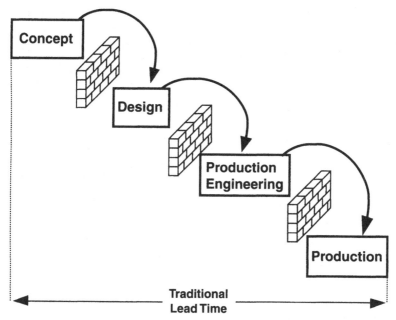

Figure 7-1 Traditional approach to process design.

Traditional Process Design

A traditional manufacturing engineering approach tosses the completed engineering design "over the wall" into production engineering, where a number of hardwired processes further solidify the profitability and efficiency of a design. Essentially, very late and very far down in the production cycle, significant decisions that affect time and money are made, which are very difficult or impossible to modify.

Often communications from one stage of the process to the next stage, from design to manufacturing, are missing. Incredibly, design engineers do not communicate with manufacturing engineers, and designers may not even know where production engineers work.

Concurrent Engineering

A newer methodology, concurrent engineering, tries to coordinate all the necessary engineering resources so that team members, working in parallel, cut down the communication time and loops involved in a new product launch. Concurrent engineering is intended to increase information sharing by bringing together all the key players in new product launches. But often these meetings happen infrequently and too late in the cycle to produce significant impact.

The concurrent engineering approach, which was pioneered by MIT's Don Clausing, represents a significant improvement on traditional, nonintegrated practice. Concurrent engineering does not, however, focus on designing specific processes to suit the product volume (takt time), and building in lean principles to meet the quality and cost requirements from the start of the launch. It also lacks the discipline to force creativity to bubble up with more ideas. And ideas are not tested through simulation before locking in a chosen process. Consequently, concurrent engineering teams may find themselves working with designs and processes well after they have been defined—once again too late in the process to take advantage of significant cost and time-saving opportunities.

Design for LeanSigma

Design for LeanSigma encompasses a methodology that literally moves across product development cycles from concept to production, as shown in the accompanying illustration. Design for LeanSigma confronts, challenges, and creates concepts and designs using a technique called managed creativity. Design for LeanSigma enables perfect quality at the lowest cost, while ensuring that products are on time to market. Very simply stated, well-trained lean practitioners can choose and apply the appropriate Design for LeanSigma tool set, no matter where a company is in the product development cycle.

Design for LeanSigma incorporates a set of technical tools that are applied from the voice of the customer through concept and design, and into production. More important, the process involves shop floor personnel, purchasing, engineering, and design professionals as a team early on. With a team focus on Lean Transformation principles, such as adherence to takt time, single-piece flow, and jidoka, the result will be improved quality and a timely launch. The best opportunities for Design for

Design for LeanSigma

Production
Process Development
Manufacture and Assembly
Concept Development
Voice of Customer

Figure 7-2 Elements of design for LeanSigma.

LeanSigma work appear whenever there are any dramatic changes in the production environment, such as:

1. New product introduction
2. Major modifications to product design
3. Major volume changes, greater than 20 percent
4. Relocation of equipment and processes

Design for LeanSigma is a powerful change process that can be used to generate new ideas or to create energy around better processes. Pella, for example, uses this design process whenever any significant production capital expenditures are proposed. Organizations like Mercedes-Benz do Brasil, Hill-Rom, Pella, and Maytag have found they can lower capital expenditure and improve product introduction with toolsets of Design for LeanSigma.

Design for LeanSigma practitioners have discovered that this process enables quicker fruition of design ideas and prototypes, and processes with much greater flexibility, enabling quick response to changing market forecasts. When it comes time to manufacture, the tools bring incredibly fast response to production operations, especially when forecasts

THE PERFECT ENGINE

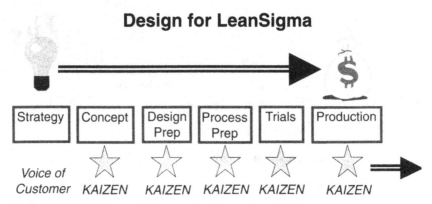

Figure 7-3 Kaizen is used from concept through production.

and customer orders significantly deviate from original projections, as they always do. Design for LeanSigma production lines focus on creating a process to produce profitability at short-term volume projections that tend to be more realistic. If your volume projections indicate the first year demand at one hundred, ramping up to five hundred by year three, the conventional wisdom will suggest building capacity and production processes to eventually produce at five hundred per year level.

Using Design for LeanSigma philosophies, however, the short-term goal is to design a process that can meet cost, quality, and capacity targets at one-hundred-per-year volume with minimum capital and cost. Now the factory can reach the five hundred units per year target by duplicating the process, with improvements at each repetition. The final quality and costs will be even better, having experienced five cycles of learning. And, if the volume projections fail to materialize, the company is not locked into a huge dead investment at five hundred per year.

A twenty-dollar bill spent two different ways illustrates this point. Imagine a long-term marketing forecast for 100 widgets a day, and the capital investment required to make these 100 widgets a day is $20. The real forecast in year one is more likely to be 10 units a day. The traditional approach is to spend the whole $20 up front on a system designed to produce 100 units a day. We say, create a design process that will produce 10 units a day and cost $2. In other words, spend a dollar at a time. As volumes increase, these processes can be added to or replicated, a

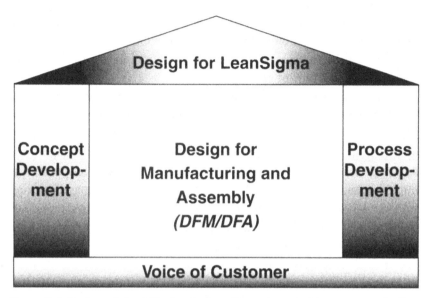

Figure 7-4 Each module of Design for LeanSigma is rooted in customer desires.

dollar at a time, to meet increased demand. That's the key to lower capital investments in the product launch process.

Begin with the Customer

Design should always begin with one key first step: gathering the voice of the customer. In a typical consulting kaizen week, team members interview customers, visit showrooms, and talk to dealers—gathering data wherever they can to begin identifying the customer requirements. Here, we discuss the importance of the open mind and avoiding the Eureka! problem.

By Eureka! we mean the sometimes natural tendency toward following one strong voice—the engineer or designer who claims, "Eureka! I've got the design." Too often, that one design will then be pursued without question. The voice of the customer can get lost in the excitement surrounding this one idea. It might be a great idea. More often, it misses the customer's desire. Another syndrome to avoid is the "Here's-the-product-where's-the-market" syndrome. Advertising and marketing

THE PERFECT ENGINE

officials know this one very well. Ever heard sales complain about the unusable bad idea that engineering dropped in their laps and said, "Sell this"? With the emphasis on teams and collaboration, Design for Lean-Sigma avoids this type of costly end-run.

Another syndrome to avoid is parts proliferation. Many companies have very creative designers, or teams that are always ready with a variation on a design that is just too good to leave on the drawing board. So another item hits manufacturing, demanding new tooling changeovers, with new issues to iron out. We want to ensure that the sheer number of variations and their associated designs don't cripple the manufacturing process. We continually encourage companies to ensure that the voice of the customer is truly driving their decisions. One of the best ways to keep control in the process, we find, is with tollgates.

Tollgates

After a team has gathered and compiled voice-of-the-customer data, it should encounter its first tollgate. This is typically a review meeting with a cross-functional group looking at customer requirements and deciding whether to go forward. Sales, marketing, engineering, finance, customer service, manufacturing, and quality should all be represented. The questions to be covered are: Can we make a product profitably to fit these customer requirements? What would be an appropriate timeline? Out of this meeting we should have target-costing objectives, project annual volume, and a product launch date.

There will also be tollgates following concept development, product design, production preparation, and trials, which are described as we continue.

Concept Development

Now it's time to translate customer requirements and develop critical design specifications. Pick up a few of your competitor's products and tear them down, analyze the design and functionality. Compare them to your existing product and voice-of-the-customer requirements and identify

Design for LeanSigma

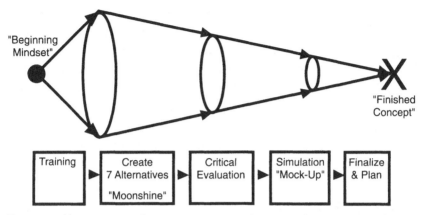

Figure 7-5 Using managed creativity, we consider every alternative in order to find the best concept.

technical specifications. Voice of the customer says what they want in simple language. This is the team's opportunity to translate customer desires into measurable features.

For large, complex products—such as construction equipment and trucks—concept development is done on many different subsystems. That means several teams must attack the development from different angles, ensuring that the voice of the customer stays true in each subsystem—the tractor's engine, the cab, and the exterior design, for example.

Concept development should also introduce the team to managed creativity. In clarifying the functional aspects of the product, we encourage teams to look to nature for the best ideas. The trick is to observe how nature solves the design problem the team faces. In one case we were recently involved in, a team was trying to discover how to suspend a motor in such a way as to minimize vibration. The team members came back talking about coconuts attached to trees and spider webs—how multiple attachment points could help stabilize a moving part.

In each case, the teams are instructed to come up with seven alternatives for each concept. This combats the Eureka! syndrome and forces creativity into the process. To choose from the seven alternatives, we evaluate ideas against technical features and specifications. Typically three alternatives are chosen to mock up as prototypes using Styrofoam, bits of

THE PERFECT ENGINE

wire, cardboard, and whatever else can be found within a few hours. The idea is to bring the three alternatives alive, in three dimensions, so that team members get a better, more physical idea of the concept. The approach allows team members to touch, see, and move the product.

As they work, the team may very well choose a piece of each of the three alternatives to go into the final prototype. At this tollgate, the team will consider the prototype, more refined voice-of-the-customer data, and better cost estimates before deciding whether to go forward.

Design for Manufacture and Assembly

With prototype in hand, the challenge now is to design a product that manufacturing can create in a timely, cost-effective fashion and ultimately meet customer requirements.

The typical first step in this process is for the team to look at the design and try to minimize the total number of parts or components. The team is forced to think about whether components need to move, relative to each other, and to be separated or attached. With the interrelated nature of each component in mind, the team can begin to minimize the total number of parts—a process that affects the cost of the product, materials selection, and labor required. Materials selection—deciding what each piece is made of—can significantly affect price and produceability, as well as quality, of the final product.

In almost every case, another round of managed creativity is called for at this point. This time the team has a more robust design to work with and several alternatives may develop for evaluation. In the design phase, each bit of functionality comes about through investigation, questioning, and decision making. If the product is a pen, the team must decide how the clip joins with the pen, whether the pen unscrews and in how many places, how many pieces the pen will comprise, whether there is a retracting mechanism, and how it might work.

After the critical evaluations are completed, the process hits another tollgate. The design may be finalized or it may be only 85 percent finished, but the company can definitely decide design specifications—dimension and function—for the product, as well as double-check the cost.

Design for LeanSigma

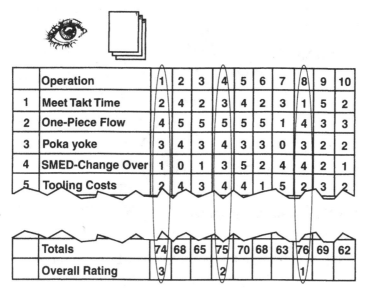

Figure 7-6 Critical evaluation process.

Process Development

A few easy guidelines determine how the team should approach product and process design: creativity before capital, quick and simple beats slow and elegant, and use resources that are immediately available. Experienced practitioners describe the creative phase of solution design as "moonshining," which is a structured innovation process that encourages unconventional, out-of-the-box thinking. We support uninhibited views—the behavior of a twelve-year-old. Look at product design and production challenges with the eyes of a child while working through simple solutions that may parallel examples taken from nature or other simple machinery.

Moonshining implies activity that takes place with some level of secrecy, in a protected atmosphere where ideas can flourish. Team members are encouraged to hold multiple creative sessions to focus on visual descriptions—sketches that connect the mind, the hand, and the eye—rather than words. For a metal assembly whose process includes casting, machining, assembly, test, and packing, sketching out the entire product

THE PERFECT ENGINE

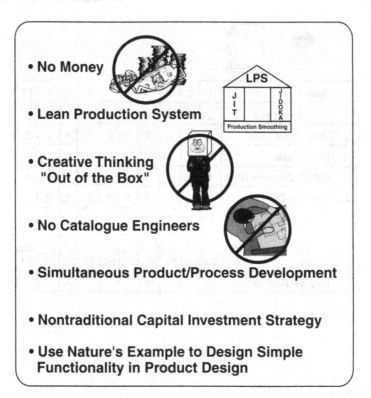

Figure 7-7 Key concepts in Design for LeanSigma.

flow forces one to think through every important detail. Changes become more real, and more considered.

Simulating real production in a mock-up cell inevitably uncovers ergonomic and cycle time opportunities. With each iteration of a good process design, team members learn how to take even more time out of the process. An entire day is usually dedicated to moonshining and the number of creative solutions every team member produces always surprises the participants. Again, for each transformation step required to put the product together, seven alternatives should be defined. Putting a hole in a piece of metal, for instance, could involve drilling, punching, machining, or casting, and each alternative should be considered and sketched.

When team members have completed their moonshining process, each

Design for LeanSigma

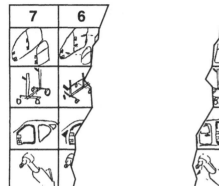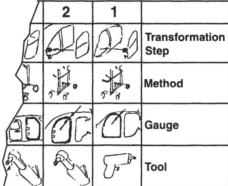

Figure 7-8 For each *process step . . . develop a minimum of seven alternatives.*

idea is evaluated and rated as to how it meets key criteria, using critical evaluation forms. The seven ideas are narrowed down to the top three.

Each of the three top solutions will be mocked up into a design or production process. The true test of every process design is how well it can be simulated or translated to a mock-up cell. Team members learn that if they cannot execute and carry forward an idea, it's best discarded.

Picking the best features from each of the three mock-ups, and then establishing the gauge, tool, and fixture requirements for each step, must be accomplished before one integrated cell mock-up can be created. The steps are laid out in a final-process-at-a-glance sheet. In a kaizen week, we put all the pieces in a cell and establish standard work.

The tollgate here will be defined by product costs, capital investment, equipment planning timeline, and trial evaluations. Process capability will also be a defining characteristic as the company decides whether to move forward.

The Owl and Grasshopper: Converting from Batch to Flow

"Getting ahead of the pitch" translates to understanding your true process capabilities well in advance of your product launch, and Design for Lean-Sigma is an opportunity to get ready for the next pitch. Every Design for

THE PERFECT ENGINE

LeanSigma experience allows practitioners to exercise their creative muscles and to combine those innovation energies with practical ideas that make lean production truly possible.

Companies like Pella, Polaroid, Mercedes, Hill-Rom, Vermeer Manufacturing, and dozens of others have found opportunity to break from long setups, big machining centers, and hardwired production lines. The process brings production associates into the process and integrates manufacturing engineers in a way that guarantees good process flows and lower equipment investments.

Further, as part of opening up the creative process, engineers and shop floor associates alike tend to develop a new awareness of the possibilities for process innovation. By learning to look at nature for alternative solutions to mechanical challenges, they are developing a new way of seeing.

A team at Hill-Rom, for example, wanted to reduce the noise of a fan that cooled electronic components under the hospital beds it produced. One engineer with fans on his mind found himself watching a Nature Channel special on birds of prey. As the day-hunting hawk comes whistling out of the sky, he noticed the hawk's noise was covered by the daytime racket of the forest or field. But the night-hunting owls descend without noise. Why? Owl wings have small serrations along the edges of their wings that deflect the airfoil, essentially producing a noiseless fan. Fan blades could be feathered, with the result that they would whirl almost silently, the team discovered, leaving the patient in bed undisturbed.

Grasshoppers are another well-recognized, naturally engineered lifting device. When planners in Mercedes-Benz were searching for less expensive capital equipment, they envisioned the strong legs of the grasshopper and found that their new press could be constructed from car jacks, a cheaper and more flexible solution to a ubiquitous mechanical challenge.

Mark Oakeson, TBM vice president and veteran lean design expert, recalls how a Mercedes team in the Brazil truck assembly plant used these methods to improve designs and prepare for production. By creating huge graphics of axle design options and plastering them over the walls for everyone to study, planners were able to reduce the design possibilities from fourteen axle designs to three. The choices were executed in Styrofoam a full

Design for LeanSigma

five months before German-headquartered engineers released detailed specs. The designs revolutionized the Mercedes design process because engineers found that they could complete process and part design before final parts specifications. As a result, when the plant's work was turned over to the German engineers, they found several opportunities for early product design changes. Oakeson feels that although there is magic in the number seven—for creation of seven alternative designs—getting to this number also wears down barriers to creativity and opens up behavior. Essentially, the iterative process encourages creativity by blending functions.

Design for LeanSigma Tips

Tip

Think about the pacemaker. Even though the product may be assembled from many small pieces, we don't want operators shuffling or moving too much. We want parts to move along on a conveyor that moves slowly at a pace equal to the takt time.

Tip

Remember, actual final assembly sequence has nothing to do with Bill-of-Material structure on the fishbone.

Tip

People need to be able to work and move without interruptions or obstructions.

Tip

Orient parts movement from right to left; use simple fixtures, rather than the body. Whenever one hand is used to hold, while the other is used to work, a fixture is needed to allow the operator to work with both hands.

THE PERFECT ENGINE

Dustbuster Cleanup

These photographs show before and after images of a Black & Decker Dustbuster assembly operation. It was a swamp of traditional high-volume assembly lines surrounded by piles of expensive raw and in-process inventory. Eighty-five assembly workers moved pieces down a long Dustbuster line, working to an unknown rhythm with little communication between associates.

This same product, thanks to production preparation simulation, has been converted to several one-person cells to produce the required volume. In each production cell one associate does all the subassembly and assembly steps to produce a complete unit.

The Walkabout, Backward

Team members were urged to study the process flow, to identify the obstacles and the bottlenecks by working backward, following the product from *finish to start*. The members were advised to simply observe, and to try to discover why Black & Decker had set up certain processes and flows.

TBM's Tom Morin, Director of Business Development, recalls spend-

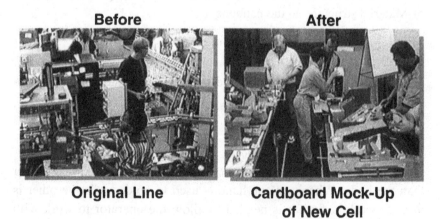

Before — Original Line
After — Cardboard Mock-Up of New Cell

Figure 7-9 A cardboard mock-up of the cell is produced.

Design for LeanSigma

ing a day and a half discussing and dissecting the Dustbuster. "We thought about fixturing; we thought about building a chaku-chaku line with fixtures that we could literally throw parts at. We ran through a number of simulations—it was an eye-opener. We learned more than we ever expected just from ten-minute observations."

The line was planned to meet customer takt time, for example, but it was running on an unhealthy five-second takt time. So for the team, the challenge became setting up multiple cells to run at sixty-second takt times. Next, recalls Morin, "We wanted each cell to have only one operator assembling the entire Dustbuster, instead of the plant's eighty-five person, two-shift operation."

Maybe there is synergy from group work or maybe there is magic in lean design work or maybe it's both, but in the end, the team designed thirteen cells, totaling thirteen people per shift. Repeated iterations of good cardboard and Styrofoam simulations in the mock-up room enabled big improvements each time the team tackled the flow.

On reflection, Morin wonders how Black & Decker ended up with its traditional, long, eighty-five-person assembly line, an example of mass production carried to an extreme with all the accompanying quality and inventory problems.

"The answer," says Tom, "lies in how they grew. The line started with half as many people. But as demand grew and people were added, the productivity dropped. As newer assembly-line workers came on, they were only adding value 40 percent of the time. The rest of their workday went to waiting, finding parts, moving materials, and endless handoffs. People were working very fast—an assembly line can be exhausting—but out of their five-second takt time, operators were adding value only two seconds. The line was completely out of balance."

Trials and focused discussions led, first pass, to a cell with two workers' worth of work. Gradually, as team members worked through more drawings and models, a better flow appeared. Says Morin, "We spent a lot of time in simulations generating new ideas."

In fact, Morin remembers a distinct change in mindset. "We had to become children with twelve-year-old minds."

THE PERFECT ENGINE

Seven Reinforcement Techniques That Build Creativity

1. Fasteners—snap-together parts are better than special screws and bolts.
2. Fixturing—some assemblies can be designed to self-fixture; look for opportunities to eliminate fixtures that hold assembly pieces as they are worked on. For example, instead of a fixture used to hold a motor in place as the enclosing casing is positioned around it, design a motor casing that acts also as a fixture.
3. No words, only pictures. When team members write words to describe a new process step, they are losing their creativity because pictures are the way the eye remembers.
4. No catalogues—creativity takes the place of catalogues offering tools and fixtures designed by outside suppliers.
5. The best kanban is no kanban. Use kanban only when you cannot link processes and feed units one at a time.
6. Jidoka—if a process is robust, operators can expect to be alerted only when it has gone out of limits or when there is an abnormality in some area of the cell. Focus on out-of-tolerance or out-of-spec conditions and assume the process is designed to run with few interruptions.
7. Learn managed creativity—use verbs to indicate action and find examples in nature to illustrate. Crocodile teeth grasp; grasshopper legs fold and spring! Draw the object, and then illustrate the specific action.

Finding the Rhythm

In every area of the Black & Decker Dustbuster cell, team members worked to create rhythm. Hands and eyes were aligned on the same plane and work was arranged so that fingers could move on the horizontal. By studying videotapes, team members worked to make every movement count and to build a cell that was agreeable to the human body. Good cell designs make repetitive work into rhythmic work, tireless ef-

Design for LeanSigma

fort that is less wearing on the body than assembly-line motions geared to serve the machine.

Polaroid, Scotland

Design for LeanSigma's international reputation is growing among companies like Polaroid and Mercedes. Over a period of twelve weeks, fifteen Polaroid associates worked together to develop a new fully integrated cell for camera production. Team members had struggled with ergonomics. The current process had workers—mostly women—seated at a high-volume, ergonomically unfriendly and extremely boring line.

During 1997, the Scottish plant was under siege by fierce competition and a threatened move to China. Workers realized there was an imminent need to reduce cost. By creating assembly cells that were linked to major subassemblies, studying logistics and labor requirements as well as quality confirmation, teams were able to implement twenty cells with tremendous productivity gains—as well as improved associate attitudes.

One woman who worked on the floor initially declared that the team would "never, ever cause her to work out of her chair." Team members persisted and told her that she was too nice to die young, and that she needed to get up out of that chair. After the redesign, that same woman proclaimed that she felt fabulous, standing up and moving. Within months, the whole factory was moving to new rhythms as chairs disappeared and new floor mats appeared overnight. The result is that product cost was reduced beyond expectations; the Scotland plant stayed open and flourishes to this day.

Pella: A New Machine

Pella Corporation is an advanced LeanSigma Transformation site that has conducted more than 1,955 kaizen events with 11,507 participants through March 1999. Lean work has decreased lead times and improved shipping performance from 89.1 percent in 1993 to 99.4 in 1999.

Better deliveries and shorter lead times allow Pella distributors to carry less stock. Distributor inventory, measured in weeks on hand, fell from 171 weeks in 1992 to 5 in 1998. Inventory of purchased parts and

THE PERFECT ENGINE

work-in-process dropped at the factories by 61 percent during the same period. Pella associates know that better process has had direct impact on profitability and market share, as well as end customer satisfaction.

Design for LeanSigma and equipment design changes play a big part in Pella's improved financial and market picture because a cleaner, less capital-intensive floor has allowed the company to bring in new products and hold down price increases for the first five years of kaizen. The associates at Pella have demonstrated unique creative strengths and steady and consistent dedication to a program that is uncommon and relatively unknown in so many other industries.

One of Pella's unique adaptations of Design for LeanSigma and the Lean Production System is the creation of their own machine build department. Pella people understand how important the right equipment is to their process, and they design their own equipment to make operators' work easier and more productive.

Why a Machine Build Department?

Equipment purchased from outside buyers, say Pella engineers, is no better than average. It should not be a surprise to see competitors working with the same machinery, and clearly that negates competitive advantage.

There is another more significant reason for internal builds of simple equipment to fit the lean process—smaller lots and one-piece flow. Historically, equipment producers have established a market for their big, batch-building machines. High-volume, high-speed runs of single pieces are their specialty, not smaller equipment made to work at takt-time pace. The equipment designed for a small work cell creates better flows and more visible operations; big equipment can get too complex for maintaining visual controls. Conveyors, racks, and other mass material storage and movement devices that accompany large-scale equipment are proven production killers.

Pella's measured and well-planned response to its capital equipment needs is a dedicated group of eighty machine builders. Sixty of those builders design their own machines; the other twenty build from corporate engineers' prints. All report to Pella's manager of continuous improvement, Brian Giddings. Giddings sets the priorities—no more jockeying for attention on plant engineering's punch list; the machine

Design for LeanSigma

Before

Cardboard Machine

After

Actual Machine

Figure 7-10 From the cardboard mock-up, the actual machine is put in place at Pella.

build team gives kaizen support its first priority. Once Pella personnel have mapped out a cell and decided on their machine designs, the twenty-four-hour team moves in quickly to execute. It's decidedly one of the strongest lean initiatives, a powerhouse foundation for lean flows.

In Figure 7-10, Pella associates simulate window sash production in cardboard. One hundred twenty days later, the redesigned sash machine is up and running, at low cost, and with fewer operators required.

Equipment Design and Build at Pella

	Before	After
Machine concept development	6 months	1 week
Time from concept to production-ready	1 year	6–9 months
Capital spent on new machines	—	About 35% less capital
Learning curve	9 months	1 month

The members of Pella's machine build department have unusual job descriptions—more of a soup-to-nuts than a highly specialized narrow

THE PERFECT ENGINE

approach. Construction crews can move and install equipment, as well as perform some welding, and run air lines and conduits. Machine builders are trained to develop all the skills required to build a machine, from estimating and design, to machining, assembly, wiring, plumbing, and even programming. They learn machining basics, as well as kaizen and other technical skills in a six-month training program, after which they join a machine build team.

"The Best Design Wins"

The Design for LeanSigma initiative is strategically on the critical path at Maytag, where fierce competition from Whirlpool and GE intensifies the company's urge to stay put in the United States and serve global markets with amazingly innovative products like the Neptune washer and the Gemini range.

Tom Briatico, Maytag Cleveland's vice president and general manager of Cooking Products, sums it up this way: "Competition is just awful. There's a price war going on. GE is doing well and Whirlpool is coming out with another Mexico-built range, a low-cost stripped-down model. The pressure is constant. We here at Maytag know that the real winner is the one that has the best design."

Against this backdrop, Briatico believes Design for LeanSigma has become more important in Maytag's transformation plan. Design for LeanSigma evolved at Cleveland in two ways. Late in the design, when the product was already tooled, associates engaged TBM to help design assembly. Then, they took the concept to design tools and equipment. Cleveland installed three standard assembly lines using the Design for LeanSigma process and found they could fix problems in advance that popped up in simulation.

Briatico notes that the approach now is to slow down a bit and use shop floor events to further perfect the actual production process. "I like this process," he says, "because if you have a lot of SKUs, using lean design to work out bugs *before* you take product to the shop floor allows time to get operators more involved."

On Gemini, Maytag's new twin oven product, teams using Design for

Design for LeanSigma

LeanSigma tools reduced final assembly labor cost by 20 percent and helped Cleveland to eliminate about 75 percent of the defects one might see in a traditional launch of an assembly operation. One Design for LeanSigma project conducted on the wiring design for a new product revealed a wiring assembly was nearly impossible to mount and took about forty-two minutes of operator time. A team reduced wiring to six minutes and made the process more operator-friendly.

Engineers have learned that because they created the design first, and the wiring harness after the fact, assembly sequence is very important to the wiring design. Engineers enjoyed the opportunity to see how these things go together and huge cost-saving ideas were developed when design engineers watched how the operator assembles the harness. "In fact," says Briatico, "design engineers have to be part of these events—they actually work in assembly. They have to set up stations with operators. We have engineers and operators working hand in hand for the first time. It's very powerful." Design for LeanSigma simulations also eliminated problems that would eventually surface in the consumer's home. Field defects that would have fallen under the one-year warranty have dropped 30 to 40 percent.

Design for LeanSigma + Design Tools = Power

Designers must understand all the elements that are critical to quality—the voice of the customer, as well as quality function deployment. "This work helps us understand lots of things that aren't manufacturing problems," says Briatico. "We need to design with as much margin for error as possible so that we can control processes. If we can loosen up the tolerances without impacting quality, we will have cheaper manufacturing process cost. So it is very important to completely understand key characteristics, tolerances and tolerance-stacking capability."

The objective is for manufacturing "right in the idea stage to help create the design, instead of engineering driving the way we manufacture. In the end, that will lead to the lowest-cost design and highest quality," Briatico says.

THE PERFECT ENGINE

Design for LeanSigma at Vermeer Manufacturing

Dan Sullivan, TBM West Managing Director, reports that Iowa equipment producer Vermeer Manufacturing now uses Design for LeanSigma to develop both the product design and the manufacturing processes for replacements to a type of earth-moving machine. A string of events, each about one month apart, has created enhancements to current product designs as team members move from concept to foam component parts.

Along the way, team members are evolving a new assembly process, almost simultaneous to actual product design. Process innovations include ergonomically assisted loading devices and fixturing and gauging for quality confirmation at each stage of the process.

Following Design for LeanSigma guidelines, team members started at the back end of the process, in final assembly, and moved through major subassembly processes, up through fabrication, welding components, painting, and "almost the complete package. It's an example of the Design for LeanSigma approach going about as deep as it has ever gone," Sullivan says.

Although this was Vermeer's first Design for LeanSigma project, results have been impressive: 70 percent reduction in labor content, a more ergonomically friendly manufacturing process, superior quality, and significant lead time shrinkage. Lead time has dropped from weeks, down to days and even hours in assembly.

Vermeer team members understand the power Design for LeanSigma holds to reduce development time and to lock in competitive advantage. They bring new product offerings to a marketplace that used to be dominated by formidable giants such as Case and Deere. Vermeer is setting new standards for customer input by including dealers in product design. And they've brought in the supply base—providers of the hydraulic system, the electrical system, even an engine supplier from Germany—because supplier expertise will help the Iowa producer understand how product will be serviced in the field.

Vermeer's new path is an opportunity to design with lean concepts from day one. Engineers and other personnel want to avoid a multitude of component parts and the myriad routings and assembly nuances that make management's job difficult. New Vermeer products must function well and be easy to put together.

Design for LeanSigma

Vermeer's Design for LeanSigma lab is busy these days with Project A and follow-on events. It's a big step outside the shop floor. Associates are pleased to learn that Design for LeanSigma is less culturally damaging than other approaches. The company has incorporated so many lean ideas into its own language—DeltaWorks, for example, has become Vermeer's Skunkworks—that executives are amazed with the speed that Design for LeanSigma has taken hold.

At Vermeer and around the globe, executives and shop floor associates alike are discovering Design for LeanSigma as a breakthrough vehicle. It makes the difference between lean improvements made long after the product launch, and total design and process control. They have seen that Design for LeanSigma is not just theory. Vermeer, Pella, Mercedes-Benz, and Maytag are all believers now in the early marriage of design and process. The power of the process has demonstrated significant results.

Design for LeanSigma is the most powerful new product design and process development tool to hit manufacturing in twenty years, a breakthrough vehicle that means the difference between lean improvements made too late and innovation from the start.

CHAPTER 8
Maintaining the Gains in a Culture of Change

Management and Perfect Process

Management is the keeper of the perfect process. Indeed it is management's role to set in motion the decisions and the mechanisms that build process strength. The role of process-centered leadership is not only to find solutions, or to develop grand plans, but also to understand and observe flows, to understand the outer limits of performance parameters, and to seize opportunities to troubleshoot and improve the process to make it robust. Truly, defects, delays, bottlenecks, and abnormalities become treasured indicators of system health and opportunities for improvement.

Raise the Bar for Competition

In fact, the moment management is *not* presented with abnormalities, it should be very clear that the process is not perfect—there may be waste in the system, or the operations may not be following standard work; there may be a host of small operational issues that need to be addressed. Further, if your new system seems to present perfect process, it's time to raise expectations a notch, because if you don't, someone else—your competition—will.

Perfection is a fragile, almost unstable condition. Every time a process

Maintaining the Gains in a Culture of Change

goes out of control, or each time a product is flawed, the process creates negative variance. Every abnormality, each time the line misses takt time, for example, represents a new opportunity for management to find out why and make the process robust.

Typically, traditional management stacks the process with allowances for abnormalities that hide the waste and fat—abnormalities like defects and down time stay hidden until they exceed the allowances or are uncovered by customers or competition. By accepting allowances in performance measurements, management is sending a subtle signal that abnormality is OK and we either don't need to or can't improve the process capability any further. It's a Catch-22: When traditional managers surround themselves with insulating layers of allowances and noncommunicating administrators and spaghetti-like communications, materials, and product flows, they prevent themselves from doing their job. Managers become control systems engineers who, upon receiving an exception signal, seek out solutions, constantly in motion to fight fires and troubleshoot.

It's a Matter of Higher Performance

But the perfect engine, a well-tuned and supremely robust machine, is one that consistently cranks out impressive operating results. Typical five-year results of the LeanSigma Transformation, for example, include these four key elements of operations improvement:

Typical Five-Year Results of LeanSigma Transformation

- 75 to 100 percent productivity gain, measured as value added per associate through waste elimination and smoother, rhythmic processes.
- More than quadrupled the inventory turns. Work-in-process inventory is usually the first one affected, typically 50 to 75 percent reduced, followed by raw material reduction, and ultimately leading to reductions in finished goods inventory.
- 50 to 100 percent sales growth through improved quality and customer service. When customers learn they will receive their orders

THE PERFECT ENGINE

on time and complete every time, the word spreads to other customers and other markets. Additionally, as the quality and the customer responsiveness improve, and are proactively leveraged, market share improves.

- 30 to 50 percent space reduction. Lean manufacturing cells require much less space for production, as well as much less floor space for inventory in various stages of conversion.

These numbers are remarkably similar to many companies' initial results. Starting out, kaizen and lean initiatives can be expected to easily reach these levels of performance. They are the results that energize the organization and enable the enthusiasm required for the long haul and more difficult projects.

But very few companies report this kind of sustained progress after twenty years of awareness. Why?

Five Years of Commitment

Five years in the life of most organizations in the Western world is a long time to sustain continued interest in best practices. There are unfortunately too many possible distractions—market swings, mergers, acquisitions, new IT (information technology) offerings, management fads, management turnover and succession issues, and even internal workforce issues—that can sideline a company that is not well and consistently managed.

Clearly, not all companies have learned the secret to sustained high performance and productivity gains. Organizations may stall at an episodic kaizen level and become distracted by other initiatives. Or they may become discouraged and drop out, or even decide that these methods just won't work for them.

It's a matter of management commitment, discipline, and focus to continually produce hard, measurable results despite painful transitions and systemic challenges. Growth always results in fundamental change that can indeed be painful, but growth is essential for a sense of security and excitement, and that is why so many organizations need to raise their level of financial and customer performance.

Maintaining the Gains in a Culture of Change

Sustaining the Gains

We recommend three essential elements to sustain the gains. We have discovered that organizations experiencing "first wins" continue to perfect their processes until their operations become consistent and predictable. This ability to continue the improvement process and to sustain the gains is based on these three critical elements:

1. Senior management leadership, including policy deployment, to set priorities and assign resources appropriate to the organization's goals; leading by example, including continuous involvement and visibility of senior management, leading the conversion to a Lean-Sigma culture
2. The Kaizen Promotion Office, management's arm—promoters, trainers, communicators—through simple operational improvement metrics, repeated and visual communication and constant feedback
3. Performance metrics that align with the policy deployment objectives are universal and simple to understand and promote the right culture and behavior

Figure 8-1 Sustaining the gain is a three-pronged approach.

THE PERFECT ENGINE

1. Senior Management Leadership

We ask all senior management leaders to promote good practices and to sustain the gains by modeling certain behavior to the rest of the organization. This will be a new challenge for most executives because it requires hands-on involvement to learn basic kaizen principles, such as work cells, Design for LeanSigma, one-piece flow, and takt time. For a minimum of one week, the entire senior management team must participate as team members in a kaizen improvement project, eating, working, struggling, and celebrating with other team members from all over the organization. It is a learning process that cannot be delegated, the first step toward organizational commitment and progress.

Next, after two to three months of organizational work in various kaizen projects, and after all senior managers have participated in a week-long kaizen event, the senior management team comes together in a three-day leadership and policy deployment session. The meeting becomes an opportunity for sharing and thinking about business challenges in a new context, because the LeanSigma Transformation is changing what associates do, how work flows, how it is measured, and how management leads. This process helps with directional alignment, brings consistency to the choices a manager makes, and promotes uniform communication of priorities.

Once each senior manager has experienced at least a week of kaizen projects by serving on a team and living the experience, he or she should be ready for the senior management leadership meeting. The event is well scripted, and each attendee has a job to do. The stated objective is to understand the company's competitive position, its strength and weakness, opportunities, and threats, and to make decisions through a systematic review of resources and priorities and a selection process that forces the group's focus on a vital few areas of high impact over the next twelve months. We call this process "Silent Brainstorming" as the participants make individual choices and set priorities from each of their own unique perspectives. Once these inputs are synthesized and articulated by small groups, in their own words, there is a strong buy-in and no one leaves in silent disagreement.

This meeting is an emotional one because it surfaces conflicting agendas, unanswered questions, and resource problems. In fact, the most difficult and gut-wrenching part of this meeting is the deselection process—deciding what *not* to do.

Senior Management Leadership Meeting Agenda

Day 1: Management Presentation	2 hours

- External environment
  - Market conditions
  - Competitive environment
  - Customer perception of the company
- Company performance
- Key business objectives
  - Performance against objectives
  - Impediments to success
- Comprehensive projects list (See Figure 8-2)
  - Benefits (qualitative, $)
  - Completion schedule
  - Resource requirement (people)

• SWOT analysis	4 hours
• Conceptual presentation	2 hours
Day 2: Establish Three- to Five-Year Strategic Direction and Growth Matrix	2 hours
Policy Deployment	6 hours

- Establish annual objectives
- Select projects/objectives
- Complete project selection

Day 3: Policy Deployment	1 hour

- Establish targets
- Establish financial impact

• Deselect nonstrategic projects	2 hours
• Establish cross-functional teams, monitoring and review process	1 hour

THE PERFECT ENGINE

Comprehensive Project List

Project #	Description	Resource (People)	Qualitative Benefit	Financial Impact	Compl. Schedule

Figure 8-2 The Comprehensive Project List will help prepare for the Senior Management Leadership meeting as well as create common understanding among participants.

Three days to conduct this exercise represent a heavy commitment on the part of management to examine, discuss, prioritize, deselect, and finally rank every possible use of company resources to focus on achieving the chosen few breakthroughs.

Outside Expert Facilitation

The exercise creates common understanding among participants, but it is important to involve an outside facilitator, at least for the first few three-day events, simply because of the surprisingly emotional discussions and political "weighting" that will emerge, elements that could pull management in the wrong direction, or into paralysis. The exercise of preparing for the meeting presentations and gathering numbers to support various activities becomes a valued annual learning experience. The senior management team prepares a two-hour presentation covering the company's activities, the external market, the internal environment, and all the initiatives that have been started.

Day One: SWOT Analysis. The team takes a slightly different approach to this well-known exercise in listing a company's strengths, weaknesses, opportunities, and threats. Instead of listing every possible strength, partic-

Maintaining the Gains in a Culture of Change

ipants are forced to chose and narrow the list down to three or four in each category. Making choices is the beginning of the process of prioritization of activities and resources; the end result should be a single, agreed-upon list of three or four strengths, weaknesses, and so forth.

Day Two allows everyone in the group to privately evaluate his or her own priorities against the company's business challenges for the next three to five years. Frequently, this is the first opportunity for managers to discover together all the various initiatives and plans they have for their groups. Once management has thoroughly examined and ranked where they are, the next step is to fast forward to where the company should be in three to four years, to identify the key measurements, talk through the lean process, and begin to map the culture change.

Directional Alignment

The first two hours of this crucial session are devoted to setting direction for the organization for future growth. The team is forced to decide whether the growth will come from related or new markets, or from re-

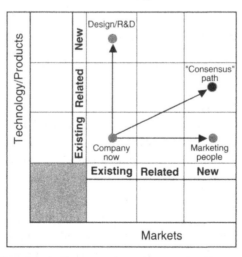

Figure 8-3 This directional alignment chart ensures that the company stays on the correct future path.

THE PERFECT ENGINE

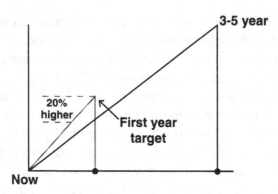

Figure 8-4 First-year targets should be 20 percent higher than the actual goal.

lated or new products, or some combination of the two. This is translated onto the growth matrix and helps to provide directional alignment and guidance for policy deployment.

Policy Deployment

The policy deployment process is focused on implementing and attaining the first twelve-month portion of the three- to five-year vision. It is advisable to make the policy deployment targets slightly higher than the linear curve would indicate—to ensure the attainment of the three- to five-year vision in spite of frequent visits from Mr. Murphy.

Lantech Experience

During one informal review meeting of Lantech management in late 1992 with co-author Sharma, attendees discovered a few disconnects as they started the policy deployment process and the directional alignment. It became clear that their individual plans for the company's future ran in different directions—for the five people in the lobby, there were five different directions. Marketing people wanted to penetrate new markets, Engineering was drooling over new products with jazzy bells and whistles, and Finance just wanted to clamp down hard on costs. These were three different agendas that could have pulled the struggling organization apart.

Maintaining the Gains in a Culture of Change

The managers agreed to conduct an offsite senior management leadership session in January 1993 that helped them reach agreement on a chosen direction, and then they completed a disciplined policy deployment with the help of co-author Sharma as facilitator. Instead of scattering focus, they decided to select a limited, controllable set of priorities, complete with action plans, schedules, and numerical objectives assigned to specific teams. And now Lantech management sets time aside every year for reevaluation of progress and new choices. It's a powerful, consensus-building exercise that Lantech uses annually.

2. The Kaizen Promotion Office (KPO)

The second powerful element that is put in place early in every successful, sustained LeanSigma Transformation is the Kaizen Promotion Office. The name is deceptive, because this is really not an office in the everyday sense of professionals clustered in cubicles, spending their time developing reports, reading or lining up meetings. The KPO is an intense collection of energetic young change agents, well-trained and extremely focused kaizen professionals "with fire in the belly" and gleam in their eyes.

These full-time lean professionals sustain the gains by continuing to schedule and organize kaizen events and followup. They are a management resource that can be quickly applied to specific problems, and they never give up. They are the promoters who implement the vision, management's arm in the daily operations.

Maytag KPO

Maytag continues to add associates to its group of KPO professionals. The job is competitively awarded, and for some manufacturing professionals, it represents a new career path. Because KPO people are bridges between traditional manufacturing processes and lean manufacturing, they become experts at design and layout of cells and new lines. They are completely comfortable drawing in many tools of the lean organization, including statistical quality control, Design for LeanSigma simulation methods, process mapping, line balancing, and forcing the organization to adhere to the standard work discipline.

THE PERFECT ENGINE

Ideal KPO candidates have fire in the belly, but at the same time they need to be a little bit naïve, not yet worn down by the crushing grind of traditional manufacturing practices and internal political conflicts. What the organization needs besides their expertise is their energy, their enthusiasm, and their ability to communicate, to exude the confidence and zeal of a kaizen warrior.

KPO associates are a special breed, the necessary assemblage of special talent that implements on the floor, teaches, communicates, and, when appropriate, feeds back issues and progress to management. As they reach into the organization at different levels, they put down roots for the new culture, which they seize and move forward. It is very important to choose the right people to grow with the KPO as the company grows in its new identity.

3. Performance Metrics

Build organizational focus and unified direction through the *metrics*.

Let the numbers lead you.
—*Dorian Shainen, Shewhart Award Winner*

Metric 1: Quality and Customer Satisfaction Index

Contrary to many "push" system concepts, all metrics start with the customer, because the objective of every system change, every internal kaizen project, every process change, is to capture and retain more customers. In the category of "Quality and Customer Satisfaction," the first metric, the Customer Satisfaction Index, is an important but neglected measure. Although many groups successfully improve their quality and delivery performance, they often neglect to quantitatively measure the impact at the customer base of all their hard work.

1.1. CUSTOMER SATISFACTION INDEX

Lantech measures its Customer Satisfaction Index level quarterly. The goal is for all its customers to be "very satisfied" on 80 percent of the

Maintaining the Gains in a Culture of Change

1.0	Quality & Customer Satisfaction	1999	2000	2001 Goal	2001 Actual
1.1	Customer Satisfaction Index				
1.2	Customer Return %				
1.3	Average First Pass Yield %				
1.4	Defects per Unit				
2.0	**Flexibility & Responsiveness**				
2.1	Delivery Compliance %				
2.2	Total Inventory Days on Hand				
2.3	Raw Material Inventory ($)				
2.4	WIP Material Inventory ($)				
2.5	Finished Goods Inventory ($)				
2.6	Average Quoted Lead Time				
2.7	Product Development Lead Time				
3.0	**Cost & Productivity**				
3.1	Value Added per Associate				
3.2	Average Labor Hour per Unit				
3.3	Average Material Cost per Unit				
4.0	**Safety & Ergonomics**				
4.1	Injuries per 100 Associates				
4.2	Medical Cost per 100 Associates				
5.0	**Financial Performance**				
5.1	Net Sales				
5.2	Operating Income as % of Sales				
5.3	R&D Cost as % of Sales				
5.4	Capital Investments as % of Sales				
5.5	Working Capital as % of Sales				

Figure 8-5 Performance measurements.

items Lantech has identified as most important to its business, the trigger points.

It is important to see clearly the link between internal changes and trigger points in your business. Given the business's competitive climate, and your customers' hot buttons, everyone in the organization should be able to tick off the trigger points on which the company absolutely must excel. For Gateway Computer, for instance, the trigger points would be delivery times, repair turnaround times, and order-taking time. For the

THE PERFECT ENGINE

Ritz Carlton, trigger points might include percentage of guests who return annually, or the number of housekeeping problem calls. For Black & Decker, they could be the number of new products introduced every quarter, and market share achievement versus expectations of each new entry.

There is more than one reason for management to focus on the Customer Satisfaction Index: first, to measure the results of all the organization's hard work, and second, to "go public," to reinforce and continue the organization's momentum.

Reinforcement and Momentum

Sooner or later most producers improve their quality, but if a company starts out from a weakened market perception of quality, do not assume that because the organization has perfected its process and made significant gains, customers will of course immediately take notice and come running. Capturing market gains from internal improvement is an iterative cycle of baseline measures, followed by improvement and tracking of progress, and finally, promotion to reinforce—or sustain—the gains.

It takes courage to promote a new level of performance to the marketplace: Going public with gains means there is no turning back. Lantech realized the impact of its improvement efforts in its market, but taking its message to distributors required some persuasion. After the company succeeded in dropping lead times from three months to a few days, it took some soul searching and a three-month delay between achieving better deliveries and going public to the customers.

Lantech's Solution

The Lantech solution—not to simply declare a lead time victory and assume market compliance—was to invite distributors into the factory, to show them what process elements had changed, and to point out the difference in the way processes had run before and after: Seeing *is* believing. Although customers may have felt that something was different, their perceptions needed to be reinforced with process reality.

Maintaining the Gains in a Culture of Change

CREATING NEW MARKETS

There is a second reason for taking the improvement message to the marketplace. A reactive position—making the changes and sitting back and waiting a year or two for the market to take notice—has the wrong impact on customers' perception of your product. When customers are assured that your product meets basic quality requirements of functionality, value, shelf life, or cosmetic attractiveness, they will inevitably focus on all the other things the organization does well—for example, how quickly phone calls are answered, or how correct the invoice is, or how useful on-line installation assistance is.

These "soft" differentiators become features of your business beyond hard product functionality, value, and cost, and they are where the winning competitor seeks and finds continued advantage. You must focus on giving your clients positive experiences in addition to quality products and services. And this is where you want your business to be, in a competitive position miles away from pure product—hardware—providers. If you don't raise the bar, your customers will.

BUILDING MOMENTUM

A third benefit from publicly acknowledging changes: As internal associates begin to hear market approval ratings and other recognition signals' feedback, their self-image and their perception of the company's operations move up. Success creates and reinforces success; motivation becomes "part of the game," and associates' feeling of power and achievement builds. Big jumps become less forbidding, and being a winner can be addictive. It's exactly the kind of positive internal attitude that all managers want to work with.

Service companies as well as hard product producers need accurate measures of customer satisfaction. TBM uses the Customer Satisfaction Index to monitor various segments of the organization, including client satisfaction, as well as new product performance. Using a 0–5 scale, with 5 being outstanding, managers can keep an eye on customer satisfaction ratings and problem areas and try to move subjective feelings about service into the more measurable area of objective rankings.

THE PERFECT ENGINE

Ultimately people make purchase and acquisition decisions based on subjective perceptions, not so much on objective data, but finding a way to capture subjective feelings is a challenge that is always tinged with the danger of skewing feedback. It is extremely important that such ratings are collected and monitored with consistency so that they provide meaningful insights for corrective action.

1.2. Customer Return Rate

The customer return rate is an indicator of the health of the process. It is an incredibly valuable measure because the data tell manufacturing so much about its own process's results in the field, as well as possible problems in packaging and shipping, or even documentation. Unfortunately, many companies must work very hard to collect this valuable data and establish a usable database. Electronics companies are pioneers in returns data acquisition and interpretation.

When a process is out of control and product is returned by customers, it is management's responsibility to investigate the cause and to use that opportunity to improve the process reliability. Essentially, when there is a defect in the process, the virus will spread to finished product, no matter how good the final inspection is.

Therefore, when the customer return rate is high, the focus must shift to internal process defects, which are measured by tracking the First-Pass Yield Rate.

1.3. First-Pass Yield Rate

If the average defect level of a factory is 10 percent, fully one-half or one-third of the defects will be shipped to the customer because even 100 percent inspection is ineffective. It is impossible to screen out all defects. The effects of this defect level are compounded if the products go through multiple steps. For instance, if in this example, there are five discrete process steps for the final product, each with 10 percent defects, the expected first-pass yield will be only 59 percent and not 90 percent.

Focus, therefore, on the average first-pass yield allows managers to go

Maintaining the Gains in a Culture of Change

to the source, improve the process, and stop the flow of defects to the customers, the *final* inspector.

1.4. DEFECTS PER UNIT

Another popular measure that is tracked effectively when producers have continuing involvement with customers after sales is the number of Defects per Unit or per Hundred Units. This measure is quite effective for high-ticket items like capital goods.

In 1992, Lantech experienced from eight to ten defects per unit, and almost 50 percent of all standard units were dead-on-arrival. It was routine to ship technicians with the product into the field in order to install and debug equipment. Four years into Lantech's lean journey, however, defects per unit dropped to .8 per unit, a 10X improvement that continues to improve. Managers seized the valuable defect information and discovered that many of the problems that had been caused by forecast errors and last-minute equipment changes were eliminated when lines started to make-to-order, building units only as needed by actual customers, to real configurations.

These product quality problems were what Lantech management called "hard defects," problems with a stuck carriage, for example, or a belt, or a nonfunctioning switch. The automotive equivalent—turning the key in a new vehicle's ignition and getting nothing, for example—represented very basic customer satisfaction issues. Today, however, Lantech and many automotive producers are happily challenged by more "fussy" complaints—waviness in the paint, or coffee stains on the pedestal, and these issues can of course only be addressed when basic product features are 100 percent reliable.

Metric 2: Flexibility and Responsiveness

Flexibility and responsiveness for a producer indicate the degree to which he can respond to demand changes in the market, as well as the changing requirements for new products and product customization. The following measures provide best indicators for improvements in this group of metrics.

THE PERFECT ENGINE

2.1. Delivery Compliance Percent

Both Pella and Maytag carefully monitor performance to customer request, within a two-day window. It is important to understand the difference between response to a customer delivery date request and any other delivery scheme, because the truly responsive producer gets product into customer hands very close to the requested date. If an order is promised—complete—for receipt on January 29, for example, and it arrives on the twenty-eighth or the twenty-ninth, the producer has met a 100 percent goal. However, an order received on January 30 would be considered late and would therefore not be included in compliance to stated objectives, and will score 0 percent of the goal.

Similarly, if the delivery is made on the twenty-ninth, but it is incomplete, the score will also be 0 percent of the goal. Tracking any measure of delivery performance, without also including performance to customer request, never improves timing within a system and builds deceptive historical data on process improvement trends.

2.2. Total Inventory, Days-on-Hand

As an indicator of lead time performance, this is the best measure that shows true overall process flow and speed. If a company turns overall inventory fifty times per year, true lead time must be one week or less. Low inventory on hand indicates financial health, generally telling us that the company is financing its business on only one week's worth of inventory. By contrast, however, four inventory turns indicate an investment of working capital on an ongoing equivalent of three months' inventory. By further segmenting the makeup of the inventory—as either finished goods, raw material, or various forms of work-in-process—we can better understand how the company and its process respond to market demands.

2.3. Raw Material Inventory, Days-on-Hand

Supply chain practices and unpredictable internal use are contributors to raw material inventory levels, although ideally order quantities should be

Maintaining the Gains in a Culture of Change

directly tied to actual customer demand. There may initially, however, be a disconnect between true demand and raw material turns as supply chain managers take a closer look at how best to buy and schedule deliveries of raw materials, such as steel and aluminum, to maintain high turns, smaller lot quantities, and good prices. The way raw materials are managed is also an indicator of how well procurement systems facilitate partnering with suppliers to improve the real value chain costs (by eliminating interface inefficiencies) and share information to improve quality, designs, and delivery of suppliers.

However, not all raw material inventory is simply due to procurement practices: It is also due to the unpredictability of internal use. Our experience shows that as much as 50 percent of raw material is there because we don't really know when we will need it. As the internal lead time shrinks and the schedule becomes predictable, this excess raw material inventory can be eliminated.

2.4. WIP (Work-in-Process) Material Inventory, Days-on-Hand

Work-in-process inventory, measured in days-on-hand, is an indicator of the leanness of a process. Inventory levels show how well one-piece flow is performing, whether the operation is running on standard work, standard WIP, the layout of cells, visibility to the process. High work-in-process inventory will be visible on a simple walk-through—crates of material waiting for processing, or in transit between operations. Tall storage racks are dead giveaways, especially if, upon a closer look, they are unlabeled or their dates are old, or the boxes are covered with dust. Supermarkets and other line-side delivery and storage mechanisms, especially when suppliers are tied directly into lines, cut work-in-process.

2.5. Finished Goods Inventory, Days-on-Hand

The amount of finished goods inventory, measured in days-on-hand, is a strong indicator of not just the good management of internal processes that are flexible and responsive, but also the level of fluctuations in the

THE PERFECT ENGINE

market demand. When management can shrink lead times and build a more predictable and reliable process, replenishment will happen faster and more frequently. Working more closely with customers to synchronize demand to true market flows improves finished goods' position.

Three competing retailers illustrate this rule—Wal-Mart, Sears, and JCPenney. Wal-Mart's success comes from moving product faster. While JCPenney and Sears maintain months of replenishment stock, weeks on hand or in transit or warehouses, the shelf life of most Wal-Mart items is measured in days, or five times the competition's turn rate. Clearly, Wal-Mart's cost of doing business is a fraction of the others', and the mix is more of what customers will buy, thus improving the customer service level.

From a "Dumbbell" to a Wire

Further, in the delivery and lead time metrics, changing the pattern of material storage (the dumbbell effect) to one of material movement through a processing operation (a wire) illustrates the flattening out of the operation and fast flows created by real-time production operations geared to market speed. Every manufacturer that takes raw material, work-in-process, and finished goods inventory out of its operation lowers the cost of doing business, and begins to focus on speed. And as manufacturers begin to resemble pure processing plants, the arch of the wire flattens to e-commerce lines.

Leaning Down the Supply Chain

At the front end of the dumbbell curve, work with suppliers for more frequent, smaller deliveries. At the back end, work with customers to improve timing. Art Byrne, CEO of Hartford-based Wiremold, a North American kaizen pioneer that has fueled its growth by acquisition through process improvement, has internally "leaned down" his supply chain. At one end, Byrne has few suppliers; now, the message to customers, some distributors, and some retailers is: Don't order on a monthly or weekly schedule—we will schedule milk runs. Every day

Maintaining the Gains in a Culture of Change

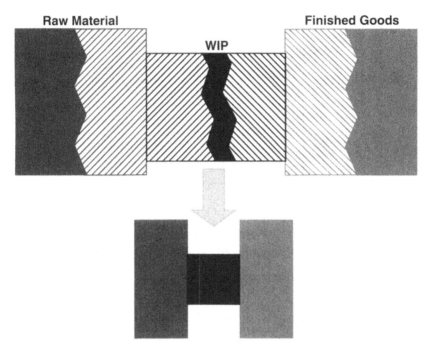

Figure 8-6 In the "dumbbell" model, extra raw material and work-in-process are required to cover unreliability in the supply chain and chaotic manufacturing schedules. Extra finished goods build up to cover unreliable replenishment and fluctuations in the market. Once a lean company develops a reliable supply chain and connects with real customer demand, expensive inventory is reduced.

you will receive your order—just tell us the night before, and we'll be there. It's a revolutionary retail concept applied to manufacturing.

A leaned-down supply chain approach is totally counter to the traditional practice of offering customer discounts to encourage higher-volume orders. It's a financially leaner and more profitable way to operate; when customers agree to take more frequent deliveries, and to work with their processors, the flow becomes almost paperless. As manufacturers that are accustomed to building to the traditional hockey-stick pattern, someone has to pay for the mound of accumulated inventory and labor costs. Someone finances inventory costs, flattening the flow with daily deliveries and inventories measured in days and hours, bending the hockey stick back to linearity. There may be a few ripples, but overall, the flow starts to look like a smooth length of wire.

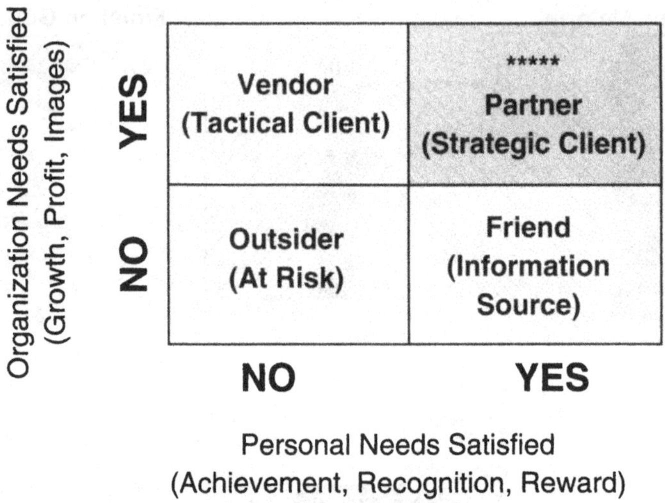

THE PERFECT ENGINE

	NO	YES
YES	Vendor (Tactical Client)	***** Partner (Strategic Client)
NO	Outsider (At Risk)	Friend (Information Source)

Organization Needs Satisfied (Growth, Profit, Images)

Personal Needs Satisfied
(Achievement, Recognition, Reward)

Figure 8-7 Customer service focus.

2.6. AVERAGE QUOTED LEAD TIME

Average quoted lead time, or the delivery lead times quoted to the marketplace, is a reflection of an organization's faith in its process. Pella worked for two years to steadily reduce lead times, but management found that distributors continued to routinely quote six to twelve weeks' lead time, despite manufacturing lead time reductions to less than a week. Finally, Pella CEO Gary Christensen and Executive Vice President Mel Haught had to take the one- to two-week lead time to the marketplace. At the same time, they worked with over eighty independent distributors to lean their processes and in some cases brought the value-adding operations back to Pella at considerable savings to the end consumer. The end result is that Pella sales are growing at a significant multiple of the industry growth.

Unipart's DCM division in the United Kingdom has applied the lean and speed concepts in supplying automotive and other critical components to after-market users for such brands as Jaguar. They have made the order receipt, picking and shipping process so lean that they guarantee to deliver parts around Europe in twenty-four hours or less, thus changing the inventory levels at the end of the value chain.

Maintaining the Gains in a Culture of Change

It's not enough to achieve big process improvements—you must take them to the market. It is an act of courage and faith that guarantees your company can and will sustain the gains. It is important to do this to increase the sales and market share so that you can provide meaningful jobs to people made available as a result of LeanSigma Transformation.

The real message behind sustaining the gains is that improvements fuel competitive growth. It's a forceful, compelling statement that the marketplace will hear: We are guaranteeing our process. It is a form of advertising your message of improved performance and confidence. The message itself also becomes a reinforcing mechanism that tells associates: We are better; look at what we can do. At TBM, management looks at the value of advertising as allocated half toward the marketplace—for customer acquisition and retention—and half for internal reinforcement and image building for corporate self-esteem. Advertising becomes a method to instill team pride, to develop a sense of belonging, to build the feeling of being number one.

Curiously, the promotion itself does not deliver and imprint the message; it's the people—the ones who created and sustained the gains—who sell it. They begin to visualize their reputation as larger than life and they work even harder, feeling self-motivated, to keep and build upon the reputation. When a crunch comes—when a retailer is all sold out, for example—that's when the process is tested and exercised. In fact, that is exactly when it is time to step up the reinforcement, to run more ads, to repeat, "We're number one."

Don't Slip Back

At the point of increased market response, the courageous producer must increase its reinforcement efforts. Pella and Hillenbrand proved this point. At Pella, as margins improved and profits started to accumulate, Christensen took a bold message to the marketplace by dropping prices on Pella's big sellers, Pro-line Windows. Similarly, Gus Hillenbrand, chairman of Hillenbrand Industries, brought in substantial incremental business for Batesville Caskets at a discount, thus creating an artificial crisis whose perceived threats caused employees to rally around his message—we can improve, and drop prices, and *still* make more money. It

THE PERFECT ENGINE

sent a powerful burst of sustained energy through Hillenbrand operations. We are not advocating price reductions in all cases, but with your newfound competitive advantage you can consider it as an option, and you can use it, where desired, to grow market share while growing profitability.

Christensen and Hillenbrand's act of faith also demonstrated the essence of leadership: to stay a few steps ahead of the crowd, anticipating and modeling desired behavior, to make achievement easier. Any medium that puts desired behavior up front—ads, books, training, Websites, recognition, awards, even an artificial crisis—will sustain the gains by setting new internal and external benchmarks.

2.7. PRODUCT DEVELOPMENT LEAD TIME

Your organization will not grow and thrive if it is simply producing the same products with an improved process. Lantech has come full circle and transformed its product development process, a typically secretive and closely guarded process, to open it up to associate involvement from other sections of the company. This transformation did not come about without a fight, however. Management remained intensely protective as new products represented the wellspring of the company. Everyone felt that the innovation process and a rapid product development process were keys to sustained growth in market share and profitability.

NEW PRODUCT DEVELOPMENT AT LANTECH: THE Q TEAM

In 1993 Lantech's development lead times were unacceptably long—up to two years to create, design, spec, debug, and prototype new models. Lead time reduction was necessary to survive and they knew it would be radical. However, Pat Lancaster, then president, reluctantly agreed to move over and try a different, team-based approach. The Q team, as the new, bigger group was called, included ten associates from the lines—welders, assemblers, and some engineers fresh from a recent shop floor lean experience. The dialogue began with a review of what the current new design was, and quickly moved into the "chaos creativity zone" from there.

Maintaining the Gains in a Culture of Change

Within one day, the same group—the Q team—that had energetically tackled kaizen on the shop floor and produced gloriously improved processes, applied their expertise to a design that had almost made it to prototype. Team members ripped the prototype design apart, reverse engineering each inch of the new model. Within one day they had more than fifty ideas for improvement. Excited by the possibilities, the Q team was challenged to slice a whole year out of development lead time. They ranked the fifty ideas by cost, feasibility, production impact, manufacturability under lean principles, and so forth. Team members were able to narrow their list down to a few good and practical ideas.

Product development's chief wasn't buying, however. "You've blown up the design—now we will never be able to make the deadline with so many changes!" He felt it would now take several additional months to make new prototypes and test them.

The only answer in this resistance-riddled atmosphere was to put down the pencils and bring in the toolboxes. By the end of the day, the team was ready to work over the weekend, and by Monday the new prototypes were ready for test—ten good ideas, modified and improved in less than three days. Bottom line: The Q team broke the lead-time barrier and introduced a new shrink wrap model one year ahead of schedule, at 20 percent lower cost, with more features. The new design caught the competition by surprise, captured overwhelming market share, and set a new standard for modular, feature-packed designs. And the Q-team design process proved that radical, intensive team-based design projects could break through development lead times the way no competitor ever dreamed.

Metric 3: Cost and Productivity

This group of metrics is designed to measure the impact of LeanSigma Transformation on the bottom line. Even though there may be a time lag in actual indication of improvements on the P&L statements, because no one should lose his or her job due to improvements, or because the old cost-accounting system hides it, we must diligently measure and maintain the operational improvements. The following three measurements are the best indicators of this improvement trend.

THE PERFECT ENGINE

3.1 Value Added per Associate

In a very lean environment, every associate's contribution adds value at some point in the process, whether the employee is an engineer, a customer service rep, a machine operator, or a shipper. The calculation that measures value added per associate highlights the financial opportunity gained by improving performance per employee without increasing head count, which is every lean manufacturer's goal.

Total sales less production, materials, and purchased services equals value added contribution. Value added, or the contribution made by the organization to the finished product, is a direct measure of the organization's effectiveness and productivity. Here is an example:

$100 million sales - $40 million raw material and $5 million plating services and subcontract machining = $55 million divided by 400 employees = $137,500 value added per associate.

The Opportunity

If sales increase 10 percent to $110 million with the same number of employees, the productivity ratio has a potential to improve significantly more than 10 percent in a truly lean environment.

$110 million sales - $44 million raw material and $5 million plating services and subcontract machining = $61 million divided by 400 employees = value added per employee of $152,500, or an 11 percent productivity gain. But the gain, as the process improves, is actually more than a simple 11 percent productivity improvement—as workers are freed, subcontracted services can be returned in-house (insourcing), and as the process continues to improve, quality and material use (defect and scrap rates decrease) to support the same sales improves. Insourcing can become a short-term indicator of productivity gains while you are waiting for sales and market share growth to take hold.

Maintaining the Gains in a Culture of Change

3.2. Average Labor Hours per Unit

At the cell and at the departmental level, this calculation is an indicator of how well associates are working. As average labor hours per unit drop—maintaining required quality levels—workers are finding the best way to process materials or to assemble them.

This is an important measure because it shows the discipline and resolve of management to take the excess people out of the process, even when sales have not grown yet. These associates become part of the resource pool to work on quality and waste reduction initiatives.

3.3. Average Material Cost per Unit

Engineering and procurement or supply chain management are key to managing material costs per unit. Measured over time, this indicator tells how well material is being sourced, used, and priced. It can be an indicator of how well material is designed into a product—what materials and sources are chosen, as well as the kind of buy procurement secures for the item. Further, payback from continuous improvement efforts carried to first-, second-, and even third-tier supplier levels will show up in this metric.

Metric 4: Safety and Ergonomics

An important part of LeanSigma Transformation is that safety always comes first, but safety is rarely measured or tracked accurately. For this reason alone we have dedicated an entire chapter to this important topic. The following two measurements can be good indicators for tracking progress.

4.1. Injuries per Hundred Associates

Companies often find that improved processes yield lowered incidence of injuries and medical costs, but they seldom measure the hard results, and sometimes it is hard to make direct payback connections between

THE PERFECT ENGINE

processes and injury statistics. Sometimes reports on compliance with OSHA guidelines may be a good indicator of progress, but we want to look beyond compliance and implement preventive measures that track any abnormal conditions that may result in reportable injuries.

4.2. Medical Costs per Hundred Associates

In 1993, Lantech uncovered some surprising findings. By comparison to six months of 1992 history, management discovered that actual cost of medical payments to associates had dropped so much that the Injuries and Claims clerk found herself improved out of her old job. Management was puzzled. "We knew we had improved ergonomics, but we didn't know the exact impact," said Ron Hicks, champion of its lean transformation. The reduced paperwork load for filing and tracking claims was the first reality check indicator.

It is important to measure this high-dollar item, especially if associates report that they are working harder, because improving processes should also improve employees' quality of life. If the numbers don't prove a positive change, human factors engineering needs to be readdressed.

Metric 5: Financial Performance

The last group of metrics measures the financial impact of the Lean-Sigma Transformation. The following five measurements are the best indicators of a positive trend. For long-term health and survival of the enterprise, sales, income, and R&D investments should grow at substantially above the industry rate of growth, while the capital investments and working capital as a percentage of sales should be substantially below industry standards.

5.1. Net Sales Growth

Net sales is a primary indicator of product and process improvement, because when product prices and lead times drop and features increase, the

Maintaining the Gains in a Culture of Change

net result, better consumer value, should be reflected in increased market share within six to eighteen months of process improvements. Certainly Pella and Lantech have noted improved sales positions, as well as profits, as each segment of their businesses has undergone transformation.

5.2. OPERATING INCOME AS PERCENTAGE OF SALES

Over the past decade, most of TBM's clients have found that after the initial standard cost-driven negative impact that can last six to eighteen months, companies experience a steady increase in operating profits that results from improved productivity, leverage, and reduced assets like inventory, space, and capital equipment—even after gaining market share. Essentially, stronger, more toned "process muscles" take the place of fat, and better body balance guarantees more output with less effort.

5.3. R&D COST AS PERCENTAGE OF SALES

Some of the income growth and the money not spent in areas such as capital equipment can be diverted to fuel a company's growth producers, like research and development. In the case of Lantech, the funds that were needed to create new products became critical to the company's survival, as competitors had produced similar equipment on the expiration of Lantech patents.

Additionally, by the use of Design for LeanSigma and kaizen methodology, Lantech was able to create a new and different type of R&D, one that produced designs in half the time, in alignment with their Lean Transformation. Lantech's less capital-intensive approach makes better use of cross-functional resources; in fact, engineers lose their Dilbertish qualities. Cracking open the development process to create faster flow taps the expertise of shop-floor personnel as well as engineers. The old image of engineers in their ivory tower cubicles, ordering from fat equipment catalogs, disappears. By building product and process design awareness and experience throughout the organization, Lantech, Pella, and Maytag have all discovered organization strength and innovation beyond line items on the capital budget.

THE PERFECT ENGINE

5.4. Capital Investment as Percentage of Sales

The LeanSigma Transformation and kaizen initiatives on the shop floor inevitably substitute creativity, human capital, and simpler machines, for heavy bricks-and-mortar capital investment and large equipment monuments. Every LeanSigma company unequivocally reports tremendous opportunities to decrease capital investment. Progress does not, however, come without conflict and challenges, because proponents of more appropriately sized equipment and work areas inevitably run into engineering's eager tendency to fall in love with machines: bigger, faster, more expensive, but not necessarily useful, monuments.

Pella's Capital Reduction

Pella, for example, achieved a 50 percent drop in expenditures for bricks-and-mortar and machinery capital investment within its first three years of lean implementation. Justifications for new equipment include the typical projections—the new machine will run faster, a new paint room will eliminate two jobs, a bigger press will cut heavier parts, and so forth. But as lean advocates have learned in every company, capital projects need very careful review and management to see if the new investment complies with lean principles—like compliance to takt time. If the takt time is sixty seconds, why buy a press that produces a part every second? Slower machines not only cost less, they also wear less.

The recommendation to critically scrutinize all capital investments against lean principles at Pella took six months of intensive persuasion at all levels of management. The final edict—"From now on . . . every request for capital equipment investments must be funneled through the KPO (Kaizen Promotion Office),"—put the decision making squarely in the hands of experts trained to watch for inappropriate investments.

Hillenbrand experienced similar capital investment challenges. A $2 million wood-drying kiln proposal project met the same fate after a shop floor kaizen demonstrated that the estimated projected capacity requirement that formed the kiln's justification evaporated. Terms like "faster," "capacity enhancement," and "cost improvement" are dead giveaways.

Maintaining the Gains in a Culture of Change

$2 MILLION TOYODA EVAPORATES AT WABTEC

Bill Kassling, chairman and CEO at Wabtec, encountered similar unnecessary capital investment requests. Having shelled out $2 Million for an automated Toyoda machining center, engineers fell in love with the big machine and they wanted another, which they proposed to use to machine large compressor housings. Kaizen facilitators disagreed, advising Kassling that with the addition of a small drill press and old fixtures, some less critical machining requirements would be offloaded and the Toyoda would be freed to machine critical parts of the compressor housing. Furthermore, by arranging this old equipment in a cell format, next to the Toyoda, the same operator was able to run the new operation. With the capacity shift, the requirement for another $2 million machine evaporated.

CREATIVITY BEFORE CAPITAL

This was a lesson in using creativity to avoid unnecessary major investments. Frequently, companies find that when they have overinvested in capital equipment, they can make big savings simply by shifting signoff and review responsibility to the KPO. The idea is to continue safety and quality and necessary maintenance expenses, but 99 percent of the requests for capacity or cost-justified improvements are removed by the simple KPO review. It may seem that management is putting handcuffs on engineers, and in a sense, that is exactly the initial intent, but as engineers and other associates become more educated to resourceful solutions to real process challenges, they turn less to big machines, and more to appropriate and creative solutions.

5.5. WORKING CAPITAL AS PERCENTAGE OF SALES

Working capital equals inventory plus receivables less payables. This important calculation is an indicator of customer and supplier relations with the organization and the relative speed of money capture in the overall financial flows. When receivables are high, for instance, typically they indicate unhappy customers, or mistakes in invoices, or less than satisfactory quality; perhaps the customer is generally unhappy with the company.

Payables are also a test of relationships with suppliers. Good relationships tend to reinforce higher payables supported by confidence in continuing deliveries and order fulfillment. Terms like COD (cash on delivery) or CBD (cash before delivery) signal lack of trust. At the Alexander Doll Company, prekaizen, assembly lived hand to mouth as suppliers demanded CBD; post-LeanSigma implementation, the company enjoys normal thirty- to forty-five-days terms.

Best of East and West Management Practices for Sustaining the Gains

In 1984, just after completing a comprehensive tour of many Japanese plants, including Toyota, co-author Anand Sharma visited a Hoxan vice president in Japan. Sharma was a vice president of operations at Union Switch and Signal, a division of American Standard, which had a joint venture with Hoxan in Hokkaido.

For three years this Hoxan vice president had worked in the United States for the joint venture, and managers from the two companies frequently exchanged visits and war stories. After dinner one night in Tokyo, Sharma asked the Hoxan vice president how he compared Japanese management practices to American approaches, now that he had experienced both. His response, "I would love to see a combination of the decision-making process in America with the execution that happens in Japan," was enlightening.

The issue, he explained, is about how to achieve managerial and organizational consensus on strategy and achieve implementation without the usual office politics.

The American Approach

The American managers are typically quick to make decisions and lay out their plans. They are pumped up and eager to move ahead. After the annual strategy session, manager A proposes an idea that is strategically aligned

Maintaining the Gains in a Culture of Change

and can move the achievement of that strategy forward quickly, and he seizes the opportunity to sell his idea to his boss. Within one day, armed with his boss's approval, A begins implementation. Manager B, filled with equally inspired and energetic ideas, seeks and also wins his boss's approval of a similar proposal, another one-day decision, followed by eager implementation. Manager C rounds out the group with his own interpretation of the best way to move the strategy forward, and armed with the boss's signoff, he begins his own implementation project. Everyone is working, scheduling meetings, planning reviews, and searching for resources.

In less than a week, A discovers that he needs help from B; the two managers start to compete, and office politics flies fast and heavy. Implementation gets ragged and expensive as competing leaders fight for valuable personnel and project resources. After about six months, results are spotty, as a few bright patches of good ideas remain amongst the debris of failed attempts and disgruntled team members. It's a damaging scenario that causes many good companies to hold off major culture change, even waiting for the next generation to come in before another attempt is made. In one week, all the decisions were made, the politics started, and within six months, only about 10 percent of the good ideas have managed to survive, each with very poor implementation.

The Japanese Approach

But in Japan, the Hoxan vice president pointed out, the process would take more time as managers worked through consensus. First, just as in most North American organizations, senior management sets a strategic direction. Manager A, who has a wonderful idea for advancing the achievement of strategy, takes his proposal to his boss. The executive listens carefully and then dispatches manager A to discuss it with each of his peers, essentially to sell (or modify) the concept. This takes time, sometimes months. But, as the idea makes its way through the organization, guided by a manager who knows he absolutely must win group consensus, it starts to spread understanding and acceptance along the way. After a few months of group selling, when the company is ready for implementation, 100 percent of the personnel understand the project, its implications, timing, resources, payoff, and rewards and risks. It's a done deal—90 percent of the exercise was in the planning and selling, the remaining 10 percent should be quick implementation.

The result is that after six months the idea is 100 percent successful and implemented.

Best of Both Worlds

The senior management leadership and the Policy Deployment process that Sharma implemented at Lantech in 1993 was a combination of the best practices of the Japanese and the American approaches to strategy implementation. Sharma used the X-type matrix that Dr. Ryuji Fukuda made famous in his book *Building Organizational Fitness, Management Methodologies for Transformation and Strategic Advantage* (Fukuda, Ryuji, Productivity Press, 1997), and combined it with the consensus-building, deselection, and powerful "Silent Brainstorming" methodology that TBM uses. The end result is that the consensus and the shared strategic direction as well as specific actions are achieved in three days and implementation success is almost 100 percent.

When Dan Burnham, now CEO of Raytheon, headed Allied Signal's Aerospace Division, he encountered a similar consensus/implementation challenge. Burnham was eager for new approaches and, wanting to "do the right thing," created a series of good new initiatives around every new idea that hit manufacturing. Lean, Six Sigma from Motorola, and Operational Excellence were each represented by a new vice president in charge of going to the plants and implementing his specialty. It's easy to understand the local reaction to these repeated incursions—conflicting agendas and bad use of critical resources stalled out many good intentions. Finally, an operations manager at the Engine Division called time: "We will do what *we* think is right," he said.

Consolidation was the answer. The Engine Division created a single Office of Continuous Improvement, not unlike the KPO; Burnham's job became one of direction-setting and strong support for change. The president and some of the staff of that division toured benchmark Japanese plants and understood their overall objectives. With the mission clear, and the resources focused under a single banner, the project moved forward.

Maintaining the Gains in a Culture of Change

The Lesson

Sustaining the gains is a matter of directing resources to the right priorities, an ideal combination of Japanese consensus building with good execution. Further, Fukuda's key to maintaining the gains was based on what we now practice as Senior Management Leadership Policy Deployment.

In Japan, Fukuda is known as the guru who focused on managerial leverage, while Taiichi Ohno worked hard on changing the shop floor. The difference between the two approaches came down to simple mechanical engineering principles of leverage—the type and amount of force applied to measurable results.

"When you make improvement on the *shop floor*," said Fukuda, "you get one-for-one improvement. And when you improve the *supervision*, you get one to three improvement, but when you improve *management*, you get one to ten, or tenfold, improvement." The methods that he believed were most effective to return this 10× leverage were facilitated by a simple, one-page deployment matrix, which he called the X-type matrix.

The X-Type Matrix

In the mid-eighties, Sony faced a problem in its global television production. Management discovered that although the company was producing units in Japan, California, Germany, the United Kingdom, and Spain, the defect rate per television unit was quite low in Japan, but it was 10× everywhere else. Sony executives believed that the clear root cause of the delta between Japan and all the others was a cultural one—Japanese workers simply knew how to make TVs better.

But Ryuji Fukuda knew better: He said the root cause was managerial ineffectiveness, not culture or geography. The brother of Mr. Morita, head of Sony, was in charge of Japanese operations, and Morita's response was to challenge Fukuda to prove his theory and solve the worldwide quality problem.

Fukuda started working at four facilities—California, Germany, United Kingdom, and Spain, using what he called his X-type matrix. Within about eighteen months, defects began to drop, as he knew they would. As a matter of fact, the defect level at the U.K. facility even dropped under the

THE PERFECT ENGINE

Japanese level. From somewhere out on the Sony production floor, he nabbed a color poster, written in Japanese: "If the English can do this, why can't we reduce the defects further?" and slapped it on co-author Sharma's desk. For Fukuda, this poster spelled vindication. He had proven that management intervention over a systematic process—not cultural or geographic issues—was the determinant of product quality.

Management Models Best Practice Behavior

Management involvement and support of this intense culture change extends beyond the blessing of the launch and periodic reviews. All senior managers will of course have learned the kaizen ways by at least one week of solid shop floor kaizen project work. It is not enough, however.

Senior management must sponsor each and every kaizen initiative. Sponsorship is more than verbal support and periodic appearances, however. Pella, for instance, includes a senior management sponsor in every kaizen project; executives are involved in selecting the project, setting the performance target, and explaining the importance of the project to the whole team. Further, senior managers must be available at team leader meetings, and for the final presentation. Systemic changes cannot be conducted in absentia.

We began this chapter on sustaining the gains with management leadership. Management must actively and continuously monitor all metrics. In fact, if metrics are not the first point to address at the beginning of staff meetings, they will be lost. The metrics must be on the agenda, up front, as a reinforcing mechanism, a measurable reflection of progress, and an indicator of trouble spots. Questions like "How are we doing?" "Are there any issues?" "How can we help?" can only be covered with the numbers behind the problems, and they must remain visibly first on the table.

Standard Work

A word about another reinforcer, standard work. We covered management metrics and modeling behavior first, as well as going public to reinforce the joy of being "number one." However, as we have suggested

Maintaining the Gains in a Culture of Change

earlier, in the Lean Production System chapter, standard work remains *the* single most influential contributor to good operating performance. Standard work is one of the most important pieces of the system designed to support perfect process. Standard work is the discipline applied to a process after it has been designed for perfect flow. Under normal conditions, it is a fragile, almost unstable condition because every time a variance appears, the process gets out of synch.

Under standard work conditions in which the process and all its elements—takt time, ergonomics, parts flow, maintenance procedures, and routines—have been established, abnormalities present themselves to management on a silver platter. Management's job then becomes one of being the problem solver—identify, eliminate the problems, and make the process robust. Standard work implies that every process can be designed and set to run perfectly—from household appliance assembly to autos and even laser eye surgery. Once the process has been mapped and redesigned, it is management's job to monitor adherence to the new process and its new standard work.

Maintaining a Customer Service Focus

Management must find creative ways to make the customer's face clear to all associates, to put the customer's presence in the heart of every operation. It becomes possible with the right data and outreach to understand the company's position from the customer perspective, and that in-depth knowledge becomes a competitive gift.

Loyal advocates are the ideal customers because they help sell products for the manufacturer. Their goodwill and public endorsement of a company's product and service causes them to come back time after time for repeat purchases based on trust and an expectation that the new model will perform as well as or better than their current one. Or, in a service business, customers' perception is that any help they need will be promptly and correctly offered—they trust the producer to provide perfect solutions.

Let's look at parcel delivery services, for example. The competition between Federal Express, Airborne, UPS, DHL, and the U.S. Postal Ser-

THE PERFECT ENGINE

vice illustrates. A customer survey would no doubt identify the top three business shippers whose volumes—but not their Customer Satisfaction Index—dominate their markets. And yet it is customer service focus that has made Federal Express the most successful delivery service. Federal Express has succeeded in putting the customer's objectives into the day-to-day work of every employee—not so at its competition. For the U.S. Postal Service to take back Federal Express's market share, it would have to become the premier package delivery system, the solution business customers automatically turn to every time they need an important package picked up. Despite the postal service's recent upgrades and bargain rates, however, customers' and possibly employees' perception of customer service satisfaction would have to rise on the diagonal, out of the "Unsatisfied," "Searching," or "At Risk" zones.

Improving the internal and external perceptions of the power of the process is the only lasting way to earn advocates. And the difference between doing kaizen activities—just to "do it"—and using kaizen as a tool to diagnose and improve a process, then sustain the gains, is understanding that manufacturing and service operations are truly a predictable management science.

When companies prove that they can identify, improve, and sustain perfect process, the response will be predictable and clear. Perfect process builds loyal, happy customers and happy, high-return customers build revenues, growth rates, market share, new product success, and profits.

Perfect process is not a one-time event. Perfect process is the fuel that drives the perfect engine, and its smooth, long-running performance can only be maintained if companies build in factors to maintain their hard-won gains—organizational commitment, leadership, the right metrics, and continued management focus. Looking into the future of manufacturing, these systemic improvements become the basis for satisfying supply chains' hyperspeeds. To build the e-commerce production model that we discuss in the next chapter, companies must take the first steps, reap the benefits, and continue to maintain the gains.

CHAPTER 9
The Value Chain

In the beginning of this transformation journey we had five inventory turns a year and now we have thirty-three. So what? Now is the time to ask, where else can we go with this Lean Transformation?

After beginning with a batch-and-queue environment, we have a lean, reliable business. As we walk through the factory we can instantly appraise the condition of our plant, capturing all that is happening in the moment. At a glance, we can see if we are making the numbers. We see that workers no longer have the mentality that they are victims of circumstance; they are in control of their own destinies and take ownership of their sphere of influence. Stop at the improvement boards and we see not only where the lines are, but also where the trends could take us.

In the business office, we see an organized workplace where each process is clearly identified. There are visible performance metrics for each function and associates see themselves as members of teams, working together. As on the shop floor, displayed performance metrics will also point the way to where the business could be going.

New products are launching ahead of deadline and the competition has been rocked back on their heels because we keep releasing innovative products with seeming ease. Our launches feature better quality than the products they replace, and we have seen a significant drop in the amount of fixed capital and working capital needed to create new products.

If we had quality defects that escaped our factory, creating warranty

THE PERFECT ENGINE

problems, we now have teams taking responsibility for continuing improvement. Not only have defects diminished, but we also have a clear idea of what causes abnormalities. Variation in the process is tracked and systematically reduced through focused work teams. We make every product every day. We are reliable. Stock outs are a thing of the past and distributors are happy.

Work teams have tool boxes now that contain all the resources they need to attack problems and to control their own destinies—from 5-S, standard work, and ergonomics to advanced statistical techniques, to concept and process development tools for new products. Teams are trained and know which tools to use when. They have learned how to act like agents of the customer, to raise their expectations of each other and of their product. Middle managers have found security in change, pride in their products and their team memberships. They view themselves and their coworkers as winners.

After much work and rethinking, we have now even become accustomed to the low levels of work in process at each workstation. The level of inventory between the receiving dock and finished goods is lower than it's ever been. In fact there is only enough to fuel the immediate work.

Now the sense of imbalance will hit home. How can we have such a great business, such pride, when our customers are not feeling the full impact of our accomplishments? Look at the chart. We have cut our lead time down to two *days* while our distribution still has ten weeks lead time and our supply chain has an eight-week lead time. Terrific improvements in the factory are insulated from the customer by an information system that is still planning and forecasting based on archaic Material Requirements Planning and scheduling systems. Standing between your customer and you are still a lot of inventory and non-value-adding activities. Between you and complete replenishment reliability stand your suppliers with their long lead times.

In the most important sense, we still do not feel the actual demand of our customers—not even a light tug on the sleeve. At this point, the only way our customer will feel the impact of our lean system is to implement a direct replenishment system based on customer demand.

When any company begins a LeanSigma Transformation, chances are that its suppliers and distributors match their lead times, week for week. If a company begins with an eight-week lead time and then works hard to

The Value Chain

	Raw Material	WIP	Finished Goods
Before	7 weeks	8 weeks	10 weeks
	?	↓	?
Now	5 weeks	2 Days	10 weeks

Figure 9-1 Our wake-up call.

lean itself down to two weeks, it has almost certainly worked selectively—opportunistically—with a few suppliers along the way. Friendly suppliers have probably been invited to participate on teams and have caught the lean bug. These are the early adopters. In most companies, however, there has been no global plan to bring all suppliers along on the journey. Most likely, the distribution system and value-add partners (for whom you are the supplier) have not even been approached about the lean initiatives.

Lean Success Stories

Manufacturers that supply retail products are especially at risk for this type of imbalance, where supplier lead times are still high and the plant becomes a lean success story. Surrounding that excellent plant, both upstream and down, are piles of inventory, extra stock that constantly faces the threat of obsolescence, especially considering today's short product life cycles. That stock requires storage, extra care, and handling and has become a financial drain.

Suddenly, the ten weeks of finished goods inventory that everyone saw as a warm, fuzzy safety blanket now looks bloated—a money pit in working capital redundancies and extra bricks and mortar. Ten weeks of finished goods inventory, and its associated high cost of materials management, is a bit ridiculous if products are being consistently, reliably produced in two

THE PERFECT ENGINE

days. Our challenge is taking the improvement techniques that created such great results inside the factory or business office to the entire value chain.

Think of it as millions of dollars' worth of finished goods sitting in warehouses: finished washers, dryers, computers, refrigerators, the kind of high-end electronics that are hot today, antiques tomorrow. All that money is tied up and waiting for customers—or theft or flood or change in consumer desires.

Here is our opportunity to extend the business model created during a LeanSigma Transformation to the entire value chain. We will address how the customer's order is received and how it is delivered. Any business at this stage will have changed dramatically—responding differently to challenges, both as individuals and as an organization. Wanting our suppliers and distributors to become equally lean and responsive is only natural. In fact, it's compelling.

More Products, Smaller Inventories

Besides, the big retailers are already pushing back on the idea of carrying big inventories. The sheer number of products we provide has proliferated. Customers want more features and options and our LeanSigma Transformation has taught us how to be responsive with customization, meaning we create more SKUs.

Now we can also grab sales opportunities. Style changes, color, and functionality changes are less problematic for the company with a lean value chain. After all, customer satisfaction can be incredibly short-term in nature, due to the competitive forces at play. A more responsive business model adds leverage, allowing us to grab opportunities as they happen. To achieve this even more effectively, we must knit together the different elements, the separate companies large and small, that compose our unique value chain.

Developing a Value Chain Vision

To start the process of value-chain-wide synchronization, we first begin with a weeklong event in which senior management creates a vision of

The Value Chain

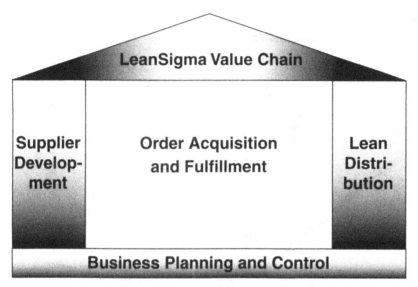

Figure 9-2 In the LeanSigma Value Chain, the twin pillars of supplier development and lean distribution are rooted in business planning and control.

what we want the value chain to be. This gathering must involve all the vested interests: general managers, purchasing, planning, distribution, logistics, manufacturing, information systems, marketing, services, and sales.

We begin by identifying the current state in a value chain map, including all information and material flows. This will be a familiar exercise for a lean company as the same techniques have been used to illustrate separate business transactions and product lines. This map will be expansive, however: beginning with receipt of a customer's order information, continuing to depict how that data is translated to create supplier orders and production plans. Physical transportation of the finished goods from shipping dock to the customer's door will be included on the map, as well as a clear chart of material components, from suppliers to the receiving dock.

This map reveals the capabilities of our value stream in terms of lead times, cost, and quality, as well as the opportunities to attack waste. In Figure 9-3 we can see there is extra inventory sitting in warehouses—both our own and our customers'—and there are three separate trucks driving for a total of seven days with our unsold finished goods. Look at

THE PERFECT ENGINE

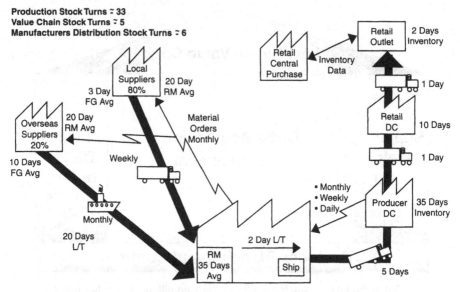

Figure 9-3 Value chain example—retail, current state.

our truck schedule. Why does it take five days to get from the dock to the distribution center? It's probably because of the way we have scheduled the drivers and trucks—or from trying to achieve isolated cost savings by only shipping full loads—but there can be a whole host of reasons that remain hidden unless explored.

Before we attack any of these issues, however, the management team must create a specific vision of the future state value chain. The future state is not the product of an individual vision. There are many ways to organize the efforts to create a shared vision, but the key is having the undivided attention and participation of senior management. We need to know where we intend to go before we decide whether to get on the bus.

Fresh Ideas, Critical Criteria

During a vision event, we brainstorm different channels and models for going to market, opening our minds to uncommon ways of doing business to spur creativity. Some ideas might be quite unconventional, but remember that sometimes great ideas can seem, at first blush, far-fetched.

The Value Chain

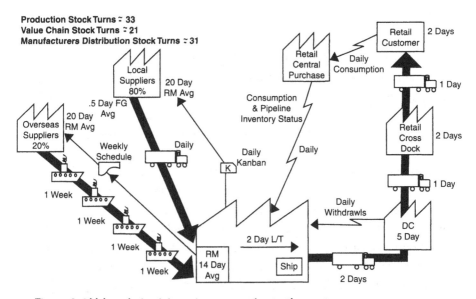

Figure 9-4 Value chain vision—increase value to the customer.

We must also establish critical criteria, such as cost, lead time, and customer service, to evaluate each idea against the elements we deem most important. Then we score each idea against the criteria and also against our own commonsense notions of what will work. From this work, we create the future state maps based on the best of this group's alternatives.

In Figure 9-4, we show a sample vision that includes inventory reduction in the distribution pipeline from fifty-four days to thirteen and a reduction in logistics lead time, from the shipping dock to the distribution center, from five days to two. Information has been rerouted so that it flows directly from the cash register to the planning system. We can see that the overall results of these changes will mean increased stock turns from five to twenty-one. By this time, we know we can go further.

Into the Gap

Next, the vision team will define the gap between the current state and the future state. Each step in the value chain is compared, present to future, beginning with the customer and ending with the customer. In

THE PERFECT ENGINE

essence, one map is laid over the other and the gap between the present and future is illustrated in the Gap Analysis.

As you can see in Figure 9-5, the Gap Analysis does not just depict the gaps between reality and the future, it also shows the cross-functional teams that will be responsible for making the transitions. For instance, the inventory in the manufacturing distribution center will be reduced from thirty-five days to five days, and the main responsibility for this part of the transformation rests on a team composed of distribution, logistics, and operations. Leadership of the project rests with distribution. The responsibilities of distribution do not end with leading this team, however. As the diagram shows, the distribution department will also support project teams in sales information, customer inventory, and logistics. This methodology ensures that the entire organization remains focused on the overall goal of value stream transformation, not solely on improvements in their own departments. Specific people need to be assigned group leadership and each team should be composed of subject specialists and middle managers.

The vision team is responsible for creating the inventory reduction goals, as well as the plan for depleting the inventory from current levels

● Project Lead   ○ Project Team Member

	What	Current	Future	Responsibility							
				Sales	OPS	Log	Pur	Fin	Dist	IT	Plng
1	Sales Information	Monthly	Daily	●	○	○			○	○	
2	Inventory Customer Distribution Center	10 Days Avg	2 days Cross Docking Operation	○		○		○	●		
3	Inventory Manufacturer's Distribution Center	35 Days	5 Days		○	○			●		○
4	Manufacturer DC Replacement	Re-order Point with Lot Sizing	Pull Based on Actual Consumption						●	○	○
5	Logistics Shipping⟶DC	5 Days	2 Days	○	●					○	○
6	Overseas (Long L/T) Suppliers	Monthly Shipment	Weekly Shipment	○	○	●					○

Figure 9-5 Gap analysis, including team assignments.

The Value Chain

to target levels. It's a decision that has far-reaching ramifications for most mature industries and requires great courage. When inventory is being slashed it will certainly have a cumulative effect, substantially reducing your replenishment demand. This reduction in demand will cause the plants to operate temporarily at underutilized levels, which can create substantial inefficiencies in operational results and, if you are publicly traded, scare the stock market. There are ways to minimize the effects, of course. We have seen companies that have successfully briefed stock analysts before embarking on this journey and not suffered a reduction in stock price. Like any worthwhile endeavor, it requires senior management planning and leadership throughout the transformation.

Following the senior management vision session, each project subgroup will independently develop detailed activity plans, including all the necessary tasks to close the individual gaps. Each task will contain goals, kaizen events, milestones, and review meetings. Project teams are responsible for creating timelines, with each task and kaizen event indicated, ensuring that appropriate resources are committed. Overall project reviews are conducted periodically to align timelines and assess results while ensuring that everyone's resource requirements are met.

Transforming Distribution

For instance, in the Manufacturer's Distribution Center where we want to reduce inventory from thirty-five days to five days, the first kaizen might be directed at reducing waste at the center. One team might attack the five-day process of receiving, putting materials away, and then picking to satisfy customer orders. A concurrent team might attack the paper transaction between picking, packing, and shipping—which has been consuming two days in this operation. The idea is to begin focusing kaizen teams on waste elimination and building flexibility into the distribution center's mainstream business practices.

If both teams are set loose in the Manufacturing Distribution Center with a goal of decreasing lead time by 50 percent—a fairly common target—that means they can cut four days from operational processing in a single week. Thirty-day homework will be required to set the new process in stone. More important, the stage is set for the big-impact projects. As part of our activity plan, we decide to attack long delivery times in the transportation system. The goal is to slash time by 50 percent each time we focus on an area, even during revisits. It takes five days to transfer product from manufacturing to the distribution center. A kaizen team will map the process, identify the delays, and focus on reducing the transport time to two days. In one instance, we helped a company with long-distance deliveries develop a kind of Pony Express system. This system focused on providing fresh drivers along the route to keep inventory moving.

Use One, Replace One

Finally, we attack the underlying assumptions that have left the finished goods sitting in the warehouse for all that time. We create a replenishment system that has a direct use-one, replace-one rule. A list of SKUs that were shipped that day would be downloaded to the manufacturing plant and put directly onto the production schedule for the following day.

Let's say it's Tuesday. Everything we ship today to the customer is downloaded to the plant. That list is compared to the production plan for Wednesday to ensure that all units consumed will be built. This is done for all distribution centers. On Wednesday, the SKUs are built. They come off the line, are loaded into trucks on Wednesday afternoon and Thursday, and begin their two-day trip. By Friday evening they are received and are back in stock. This cycle repeats every day.

Selling the System

We also need to approach our retail customers and tell them of our newfound capabilities in the plant and our distribution system, emphasizing our speed and reliability. To our customers, this means they no longer need to carry redundant inventories and can reap the benefits of immediate cash on the balance sheet. Before our customer reduces stock, however, we will certainly have to prove ourselves. This will take time as the retailer learns to trust our ability to deliver. At this point, we're going to send a kaizen team to the customers' warehouses and work with them to set up a replenishment system, modeled after our own replenishment sys-

The Value Chain

tem, built on the principle of take one, order one. This will link directly to shipments from our own distribution system. It is our hope and expectation that down the road, our customers will no longer see the value of maintaining their own distribution center, thereby creating a direct link to the retail outlets. This allows us to resupply directly to the store. At this point, we are monitoring consumer use daily, looking for changes in mix and volume and using those numbers to drive our demand planning.

Supply-Side Transformation

After all of this valuable work in the distribution system, however, we are still lopsided and vulnerable unless we pay close attention to the supply chain. The first job of supply-side projects, which was accomplished by the executive vision team, was to create a current map of the supply chain and select key suppliers. To select suppliers, our purchasing executive has shared with the subteam which purchased parts—or which suppliers—have delivery reliability issues, excess inventory, or quality and cost issues. The decision is important because most companies have hundreds of suppliers large and small and we need to begin by selecting a few with a lot of leverage in order to ensure success.

With these suppliers in mind, the vision team has already drawn a future state map for the supply chain and created a Gap Analysis. So we know what the important tasks are, but the next step should be considered carefully as the situation is somewhat delicate.

With one major automobile company, we discovered several years ago the first steps *not* to take. That company selected suppliers and simply exerted pressure, telling each company that it needed to start doing this kaizen thing—like GM's Lopez-supplier-squeeze approach. The suppliers were told, with no prior notice or preparation, that TBM consultants would arrive on a certain date in their business and they were to make time, pay the consulting bills, and, essentially, do what they were told. With little explanation, there was little buy-in.

We have seen smaller organizations very successfully draw key suppliers in to lean improvement strategies simply by inviting top executives to participate on kaizen events. The kaizen events should directly pertain to the

THE PERFECT ENGINE

supplier's area, of course, and the invitee should always be top management decision makers, not salespeople or midlevel engineers. During such participation, we have seen dramatic improvements in supplier quality, delivery, and even design of components to meet customer needs. Quite frequently, these experiences by supplier executives result in the supplier's own lean improvement efforts. In the end, everyone wins.

A Better Model

Larger businesses with hundreds of key suppliers could follow the model we established with Chrysler, to some degree. First and most important, Chrysler implemented lean within its own plants. This is a prerequisite. Then three or four top executives from perhaps twenty key suppliers were invited to one introductory meeting at which the head of purchasing explained the cooperative spirit of the new program and TBM executives outlined the Lean Production System and the requirements for the partnership being extended. Chrysler would pay the cost of consulting and training, suppliers were told, for suppliers interested in splitting the benefits from improvements with Chrysler, fifty-fifty. Those who stepped forward and volunteered—as many of the attendees did—were then set onto a schedule for a personal kaizen breakthrough experience.

Knowing the improvements that Chrysler wanted from each supplier, TBM conducted awareness training for the supplier executives, covering the Lean Production System. At a later point, kaizen weeks were conducted on the supplier's site on key product lines for Chrysler. The improvements Chrysler realized were tremendous, but they could have gone even further with a more organized Value Chain project.

Now we would recommend complete openness with suppliers about the current and future state maps. Each key supplier would be given a copy of its own Gap Analysis and challenged to close its gap. We would set up a schedule for kaizen events in its plant and welcome the supplier at strategic planning sessions to discuss the progress being made. Ultimately the goal is to install a system for replenishment that is part of the same system implemented with distributors—a pull system. Building on principles of "use one, build one," work with suppliers focuses on achieving the necessary flexibility and response triggers so that the system provides a reliable flow of materials. Next, everyone must then move on and tackle the most persistent issue to plague a value chain, a department, or any relationship: communication.

The Information Challenge

It is important to consider information, information systems, and their role in facilitating communication in order to ensure that all functions and value chain partners are aligned in support of the customer. A lean value chain can only survive and be effective when information flows directly and freely. Shared information is traditionally distributed selectively, in batches. This is where foxhole management thrives—where information is power and data is jealously guarded. These old-fashioned responses cannot be allowed to continue.

Outmoded information systems like MRP and ERP won't do either. The biggest problem with these systems is their complexity: They mirror the tangled web of old-fashioned functional manufacturing. Now that our processes have been streamlined and simplified, these archaic systems are no longer relevant for our new reality. Rather than scheduling every single operation as we did in the old world, we only schedule final assembly. With a visual pull system, all supporting operations are linked to final assembly—including the supply chain. The future role of MRP might be what it was originally designed for: to give suppliers a long-range material requirements outlook so that they can plan for their business.

The important thing to realize is that we are carrying far less inventory in the system. Information becomes our substitute for inventory. We used to hold stock in order to respond quickly, because our lead times were so long; now we have abbreviated our lead times and timely information is crucial. And we must share customer pull information across the value chain. Customer and value chain partners cannot be supported without a free flow of information.

Business Planning and Control

All of our value chain improvements are real, but they will not be sustainable until there is a system of moving facts and decisions between in-

THE PERFECT ENGINE

dividual departments and value chain partners. The effectiveness of planning is firmly rooted in the quality and timeliness of information and the competence of decision makers.

This free-flowing record of our customers' constantly changing buying patterns must be funneled into the planning process. That data becomes part of decision making for realigning and redefining resource commitments so that the extended organization operates in a synchronized fashion.

Marketing receives point-of-sale information, for instance, once a month—a planning cycle that is appropriate for most people. They analyze it to determine changes in customer buying patterns, product by product. After the information has been analyzed and compared to the original idea of what was going to happen, changes to the plan are made and shared with manufacturing before the planning becomes fixed. At a monthly meeting, the sales plan, production plan, and supply plan for the current month and beyond, plus the resulting inventories are adjusted and agreed upon. In this scenario, nobody lobs grenades because everyone has so much information going into those meetings that there are few surprises. A contract should also be established here between sales and production, laying out everyone's expectations. The monthly financial plan is the result of this contract.

Information flow should be constant. The relationship between actual and planned sales is constantly monitored. As trends emerge, up or down, they are immediately shared with production planning so that timely contingencies can be developed in support of the customer. Production planning must then share this information with the supply base so that suppliers can take action in their own facilities. While this appears to be a reactionary process, it is based on the philosophy of sell one, build one that is the new foundation of our value chain.

Recreational Vehicles on the Value Chain

Maytag has been using LeanSigma and kaizen methodology to perfect its basic production and supplier delivery processes. Now Maytag is extending its new capabilities out through the value chain, product line by product line, instead of waiting for the entire company to catch up.

For instance, Maytag's Cleveland, Tennessee, plant is the major supplier

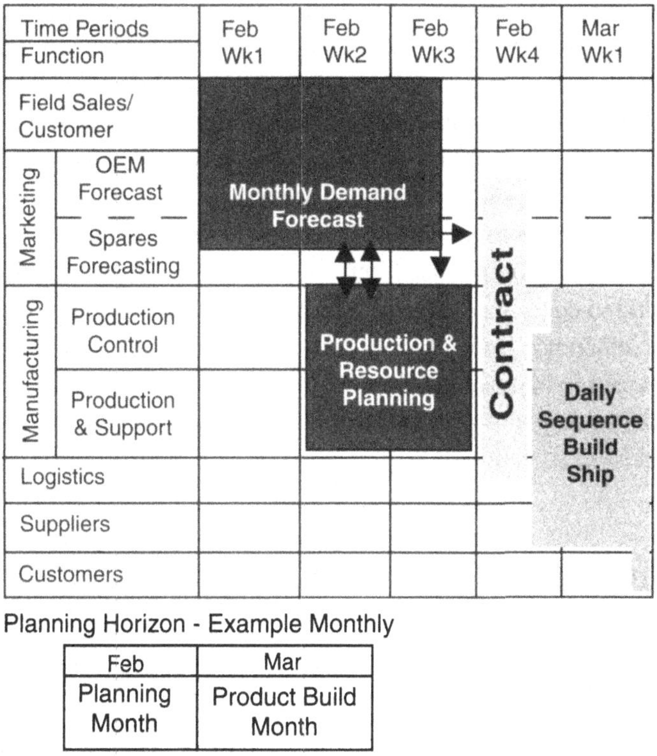

The Value Chain

Time Periods / Function		Feb Wk1	Feb Wk2	Feb Wk3	Feb Wk4	Mar Wk1
Field Sales/ Customer						
Marketing	OEM Forecast		Monthly Demand Forecast			
	Spares Forecasting					
Manufacturing	Production Control			Production & Resource Planning	Contract	Daily Sequence Build Ship
	Production & Support					
Logistics						
Suppliers						
Customers						

Planning Horizon - Example Monthly

Feb	Mar
Planning Month	Product Build Month

Figure 9-6 Monthly planning activities.

of ranges for recreational vehicles. The line represented a good starting point for beginning the long-term initiative of creating a build-to-replenish system. We began by creating a detailed value chain map of the current state, depicting each step between Maytag's plant where the ranges are built and an RV plant in Indiana—the RV production capital of the nation—where the ranges are installed in galley kitchens. We found redundant inventories. There were seventeen days of inventory in a warehouse in Cleveland. Ranges sat for another thirty days in an Indiana warehouse. It was abundantly clear there was a big opportunity for improvement.

Maytag was already flexible, already capable of making every product every day on the RV range line. The question was, how to extend this model line's capabilities along the value chain. An interesting subplot or side benefit emerged as manufacturing perfected its processes and cut inventories.

251

THE PERFECT ENGINE

Although the RV assembly line had been locally thought of as one of Cleveland's most capable, it had a wider reputation for being nonresponsive, with long lead times and disappointing schedule compliance. Improved production performance to actual demand helped Cleveland clean up its act. Production understood that it needed to hardwire better schedule adherence and service to customers.

The first step was to go back to the line with the goal of achieving 100 percent schedule compliance every day. We discovered that we really just needed to correct a couple of internal supply issues. Within ten days, schedule adherence jumped to 99 percent and attitudes changed as associates learned to balance their schedules daily instead of weekly. Operators on the line rose to the challenge and the swiftness of their response exceeded all expectations.

A New Beginning

The next step was creating a kaizen team with people from distribution, operations, systems, purchasing, and marketing with a goal of establishing the mechanisms to link a distribution center in Indiana with the production line in Tennessee. A buy-one, make-one system was created with immediate pull to the plant—no more side trips through corporate offices. Whatever products were used in the distribution center were listed and downloaded to Cleveland and became part of the build schedule the following day.

The team identified six key suppliers to the RV range manufacturing process and linked them to a direct pull system. The system fed component usage information from the line to all suppliers several times a day. Ryerson Steel was already on a simple kanban pull system, so going to a direct pull system to convey usage and target fill levels was not difficult. Suppliers now love Maytag's kanban system because it allows them to deliver directly to shop floor supermarket racks and availability improves immediately. The system was extended to inside suppliers, such as the fabrication shop.

Software issues and other implementation questions emerged second in the list of priorities for Maytag. The idea was to enable a perfect process, bring in suppliers, and do it in a simple, clear way that allowed for constant improvement—with no hardwired monster solutions.

The Value Chain

Building Faith and Trusting the Process

Running the system to a real replenishment signal, rather than a forecast, is a big change for Cleveland, as well as headquarters planners in Newton, Iowa. The change tests the faith of headquarters in Cleveland's production promises and the supply chain's ability to be consistent and in control. Headquarters in Newton is still running marketing and some inventory and capacity planning, but daily production schedules are set to respond directly to customer needs. "That," explains Ramin Zarrabi, Maytag's Value Chain project leader, "is how we went from thirty-seven days' finished goods inventories down to fourteen days in six months."

Maintaining the Gains

Without discipline and measured results, sometimes operations finds it easy to slip back into traditional practices. Maytag's model lean value chain is reinforced with metrics that maintain the reliability of the supply process and keep inventories low.

Now that Maytag has experienced the power of a lean value chain, it is

---

Benefits
Consumer, Customer, and Maytag Value Proposition

- Improved customer responsiveness
- Ability to personalize orders by consumers and/or end users
- Process simplification
- Lead time reduction from order to delivery (customer and Maytag)
- Quality improvement
- Finished goods inventory reduction (customer and Maytag)
- Productivity improvement
- Transforms Maytag from push to pull culture
- Improved consumer/customer satisfaction

Figure 9-7 Benefits accumulate throughout the transformed Web-enabled enterprise.

THE PERFECT ENGINE

making plans to extend these techniques to other product lines within Cleveland, Tennessee, and beyond, to other corporate brands.

Extending the Gains

"The future is pure build-to-order for customers like Sears, Lowe's," says Zarrabi. "We will enable these customers to place an order and expect same- or next-day ship."

It's a logical progression, say Maytag planners, to move from traditional practice to build-to-order in days. The company recognizes a need to reduce finished goods inventory, build to replenish, then build to order, one product line at a time.

Much has been written about how consumers have taken control of the buying and selling process. Nobody waits for advertisers to tell us what we want anymore. We bypass marketing. We skip right past the local shopkeepers' available inventories to find the distributor who has what we want. We are no longer willing to wait. The Internet age is upon us.

The traditional approach has been to accumulate inventory in order to be responsive to this impatient customer. That has been the downfall of

---

E-Linkage Metrics

- Customer Satisfaction Index (Voice of the Customer)

- Daily Schedule Attainment

- Finished Goods Inventory

- Total Cycle Time (Order to Delivery) Reduction

- Damage Claims Reduction

Figure 9-8 Metrics used by Maytag to maintain and reinforce the e-process.

The Value Chain

more than one dot-com company that has drowned under bloated, expensive inventories and brick-and-mortar warehouses, and armies of people to receive, stock, and control all that material. The business model does not work because of the competitive pressures on price. Nobody can afford all of that infrastructure for long.

We have described here a process that allows companies to offer excellent customer service without the need for expensive warehousing and inventories. The Dell model is a good example of a build-to-order system. Order a computer over the telephone or on Dell's Website, completely customized, and Dell assembles the order the following day and ships it directly to you. Business models built on lean principles like this, extended through the value chain, will provide the competitive edge for those companies that take the initiative.

CHAPTER 10
The Future

The present tense does seem oversized at times. This new century has proved to be an era of massive corporate mergers, international acquisitions, and dizzying stock-market swings. The only way to survive, it sometimes appears, is to eat all the other fish in the pond.

When we look to a brighter future, however, we see something much smaller. We see manufacturers splitting up their multimillion dollar factories in favor of discrete assembly units positioned in the same neighborhood as a majority of their best customers. We see proximity creating new personal relationships that lead to better understanding of customer needs and, ultimately, to better product development. We see customer loyalty built on delightful and responsive service experiences—allegiances built on face-to-face contact. In short, we see Distributive Manufacturing.

Distributive Manufacturing

This is an idea that frightens some business executives who fall into the trap of believing that decentralization means losing control. After all, employees are scattered across the nation or globe. It is no longer possible to simply take a walk and lay hands on each process. As we sketch out a future built on Distributive Manufacturing, however, we ask that you keep posing this question: Which relationship is most valuable to your busi-

The Future

ness, the one you have with your customer or the one you have with your colleague?

If you were a supply-side business, just how valuable would it be to your business if you visited your customers' assembly lines twice a day? Many final-product producers—who have spent too many years outsourcing assembly and components to foreign countries—have already discovered the benefits of abbreviating lead times by sticking close to their customers. Truly progressive companies are now looking to the time when their order-to-ship cycle is five days or less. What we are talking about is creating a full order-to-installation cycle that is five days or less. To do that, you must be close to the action.

In the future, we envision smaller plants located in and designed to produce for local markets, creating exactly what local consumers want, when they want it, without many insulating layers of marketing, logistics, or materials management in between. It's a radical change from traditional patterns of production, and even from transitional systems such as Dell's and Gateway's custom build-to-order, Web-enabled approach.

From the beginning of the Industrial Revolution, new technologies—from the steam engine to the telephone and computer—have always driven business into new arenas. And there have always been a canny few who see further, into what new technologies will allow. Communications technology developed over the past few years, for instance, certainly make the decentralized future easier to attain. Instant communication between headquarters and discrete business units, via the Internet, can ensure that daily operations remain aligned with company-wide objectives. With Web-based video links, engineers and executives will be able to gauge the effectiveness of remote operations in real time. But this is not just about the Internet. Even if all communication is done by fax or well-trained carrier pigeon, the tenets of Distributive Manufacturing remain focused on responsiveness to the customer—as well as responsibility to the community and quality-of-life issues for associates.

Distributing Operations at Toyota

Where Distributive Manufacturing is the most intuitively right is in big-cube products. Shipping large items across continents is not only costly,

257

THE PERFECT ENGINE

it's slow. Your production lines might be waste-free, your process might be perfect, but if you cannot be as responsive as the hometown producer, you may have still lost the battle.

Responsiveness—shorter lead times, reduced shipping—is a good reason to shorten the distance between your operations and your customer. But the best reason, which Toyota has been continually discovering since the early 1980s, is goodwill.

Cars have a particular place in the American psyche. It's not just that they are expensive items, which most families do not often replace; they are also illustrations of our personalities. The man who parks his big, shiny truck in his driveway makes immediate assumptions about his new neighbor just emerging from a battered sedan. It wasn't a surprise, then, that in the years following World War II, Americans would have thought themselves traitors to drive a Japanese car. Even after Toyota emerged from the ashes and built cars that were obviously superior, even forty years after the end of the war, the United States resisted—staying away from Toyota distributorships in droves.

In 1973, trying to get closer to its customers, Toyota opened a research and development facility in California. Living among us, the car company was better able to judge what we needed and wanted. But it didn't change *our* feelings toward *them*. Changing the mind of the average citizen didn't happen until Toyota made the much-heralded decision to open a production plant outside Oakland, California, and begin using American workers—and American suppliers—to build Corollas. Even then, in 1986, there was debate about what Toyota meant by this. How much was it really investing? Some portions of the car were still being produced in Japan. Did that mean they didn't trust us? Newspapers carried speculative stories for months, particularly in the Oakland Bay Area, as we watched the facilities going up, the people being hired, the suppliers getting contracts. Money flowed into the community. Toyota employees played in softball leagues.

After a time, Toyota was not a Japanese company. They were us. Relatives worked for Toyota or for the new industries that had sprung up around the factory. We were them. In less than a decade, Toyota was the top selling car in the United States.

Look at the Toyota Website now and you can see the words and pic-

The Future

tures offer one constant message: We are you. The carmaker points out that Toyota purchased $13 billion worth of parts and materials from five hundred North American suppliers in 1999 alone. More than 75 percent of vehicle content is now purchased from North American suppliers; four North American vehicle assembly plants produced more than one million vehicles that same year. Toyota lives among us, studies our desires, hires our children, and services our cars. How can such a company be considered foreign?

True, Toyota has not always had an easy ride in the past fifteen years. The Japanese recession in the early 1990s hit Toyota hard and market share dipped. But the company was at least partially insulated in its U.S. operations by the fact that it could buy and sell in the same currency, unaffected by the yen's fluctuations. For any company, being able to produce and sell within a single currency is a benefit of Distributive Manufacturing. There are enough surprises for an international manufacturer—from cultural clashes to supplier and shipping problems—without getting the balance sheet caught between fluctuating currencies.

Build-to-Order

Just recently, a colleague decided to order new shades for her living room out of a catalogue. It was a very nice catalogue, offering hundreds of choices of fabric, hem types, hardware, and style. To her delight she found the company had a Website where she could instantly order the shades so that they, according to the Website, would begin being fashioned the following morning. Excited to have found such an obviously lean and responsive company, she put in her order only to find that her shades would not arrive for at least eleven days because the manufacturing location was 3,000 miles away. Delight turned to disappointment.

What a wasted opportunity. If that manufacturer, camped in one location on a distant coast, could spread its employees and sewing machines into small shops in selected regions of the country—the regions where most new home building was occurring, for example—it could have offered to deliver new shades to our colleague by the weekend. Taken one

THE PERFECT ENGINE

step further, the company could have offered to install the shades and transform the woman's home. How loyal do you think the customer would have been then?

Assembling sewing machines and fabric to make shades is a pretty simple example. A new shop could be set up in a rented industrial-park space and fulfill orders within a week. What about more complex operations?

Rethink Process Design for Distributive Manufacturing

For many companies, the sticking point to imagining dozens of smaller assembly posts is the painting and finishing operations. The same paint booth that becomes a bottleneck in the factory would seem to be an anchor on companies that cannot imagine replicating their paint booths all across the country. The time and money spent on getting environmental variances for paint booths from local governments would be prohibitive. There are ways, however, to cut the red tape.

Consider our friends at the Batesville Casket Company. This is the world's leading producer of burial caskets, headquartered in Indiana—the middle of North America. It's an incredible business with enormous shipping challenges. Even with manufacturing facilities in Indiana, New Hampshire, Mississippi, Tennessee, Quebec, Canada, and Mexico City, there are still corners of the globe that cannot benefit from Batesville's lean responsiveness.

But if Batesville prefinished all of its materials in preexisting manufacturing facilities and then shipped out the flat materials unassembled to smaller, more flexible operations for converting into finished caskets, imagine the possibilities. If the company saw, for instance, that Texans were more partial to high-gloss mahogany with gold-tone fixtures, while Floridians preferred metallic finishes, it could target its market and reduce finished inventory. With Distributive Manufacturing, there's a much better chance of the local assembly operation becoming intimately familiar with the preferences and needs of the local market. And if you could promise the mortuary twenty-four-hour delivery or less on any type of casket ordered, customer loyalty would skyrocket. Further still, if they could customize or personalize each casket for each customer without additional lead time, customer satisfaction would skyrocket.

The Future

The Electronics Industry

The electronics industry is typically ahead of the curve when it comes to fast production and rapid-fire introduction of short-shelf-life new products. Such companies as Dell, Gateway, IBM, and Winbook have pioneered faster approaches to order fulfillment with quick builds and cash realization even before delivery, facilitated by Web-based systems. Although none of these pioneers have succeeded in wresting complete control of the supply base, the electronics winners are pioneering lean manufacturing and responsiveness. And there are lessons here about localizing lean, demand-driven production centers right in the heart of the consumer's home territory that other lean manufacturers need to observe and understand.

Denver's EFTC, for instance, has taken entire links out of the supply chain for its big Japanese computer producer, Fujitsu, by locating Tennessee assembly operations next to its logistics provider. Basically, such smaller outsourced "subcontract" operations as EFTC can provide quicker, more responsive localized production. Headquarters creates strategic direction, but the daily tactical decision making is handled easily on a local basis.

Operations such as automotive, appliance, consumer electronics, and even customized clothing, which were shipped offshore for a supposed global price advantage, will come back to local consumers because lean processes and responsive manufacturing systems will prevail. Design for LeanSigma, developing efficient and innovative products and appropriate manufacturing processes that are tied directly to actual demand, is a critical piece contributing to a Distributive Manufacturing reality.

Enter the Internet

The Internet has flattened the value chain by shifting more power to the end consumer and making most intermediaries redundant. Now, the map of the typical value chain enterprise looks simpler, with fewer interruptions in the flow and fewer solid lines connecting customers to warehouses and truckers, subassembly plants, and suppliers. Eventually the map will basically string together the customer, a lean supplier, a direct delivery system, and perhaps one or two tiers of suppliers. The winners understand that their survival and prosperity will depend on leveraging

THE PERFECT ENGINE

manufacturing capabilities to respond first with perfect process and holistic solutions, right in the backyard of their customers.

What Will This Mean to the Workforce?

Much of what manufacturing and supply chain leaders have accomplished in the last 150 years has been done through human hands, but without the full engagement of human minds. The LeanSigma Transformation addresses that failure to recognize and position human capital—the only appreciating asset an enterprise has—at the very front of a transformation and offers structure to more fully realize human potential. Distributive Manufacturing, however, can take this a step further. Although it is impossible to predict exactly when an envisioned and empowered, innovative workforce will run distributed manufacturing sites, understanding their desired characteristics is not difficult.

If executives follow the simple guidelines given here, the associates who design and operate the new Distributive Manufacturing systems will be well-prepared, trained, and disciplined, and will be recognized and rewarded for their valued contributions. It will be impossible for the leaders and operators in the new Distributive Manufacturing plants to remain neutral or uncommitted. The very visible transfer of power into the hands of people directly responsible for adding value to the company's products and services raises the bar for accountability, results, and inspiration. Expect to witness the departure of workers (and executives) who love routine, and whose expectation is that someone else will solve the problems. Running a good operation becomes the collective responsibility and pride of teams of trained professionals; we can expect that shift to be a remarkable and perhaps difficult one.

Good processes, the fuel of a Perfect Engine, create wealth, and wealth creates growth opportunities for empowered workers. It's an incredible opportunity for manufacturing and supply chain professionals to lead a new Industrial Revolution. The next decade in manufacturing will continue to be filled with great excitement, some battles, streams of high-energy innovations, and the arrival of newer, better-prepared leaders. Intelligence will return to our processes, and we envision growth and businesses leveraged solely on the successes of manufacturing operations.

The Future

Last Words of Caution

Sending a cross-country traveler off on the road with a full tank of gas, a cooler loaded with sandwiches and water, a flashlight, pencil, good map, and a cell phone is not a bad way to think about this LeanSigma Transformation journey. We have packed all the tools you will need to plan your trip and get started. We have loaded a map of the region, a flashlight for emergencies in unknown areas, and the cell phone to stay in touch. The only travel aid we cannot supply is a certain sense of adventure and energized eagerness for the journey and for the discoveries to begin and, of course, your personalized vision of the future.

It's possible that your company may start the ignition and pull away into the fast lane immediately. Or you may choose to take a more scenic route, stopping along the way to reflect on your progress and to enjoy the experience.

It will be a good trip, filled with surprises and hard work and very visible rewards. The landscape will pass by quickly, and you will be amazed at how rapidly you pass from one time zone to another. You may pick up new passengers along the way, and ideally everyone will arrive at the rest stops together, refreshed and eager for more adventure.

But just in case, we want to pass along a few words of caution for the trip:

- There is no need to bring everything for the trip. You will acquire the things you need as you go. Some items that were packed in anticipation of great need will be thrown out the window—such as battery-powered portable translators, deer whistles to mount on the bumpers, books on ERP, and strategic planning tomes. Bring the essentials and pick up what you need along the way.
- Do four hundred miles every day before stopping. Without a set number in mind, travelers sometimes linger at lunch, or oversleep their departure times.
- At the end of each evening, review your progress—what did you see? Where are you headed tomorrow? Is everyone ready?
- Plan rest stops—short breaks to walk, think, and meet with colleagues and friends are more effective than long stretches driving into the night, followed by hours of recuperation. No scenery, much boredom, and no progress take the fun out of the ride.

THE PERFECT ENGINE

- Be alert! Watch for climate shifts and markers indicating changes in the terrain—hardwoods giving way to evergreens and rocky cliffs, or open fields bordered by small creeks. The passing landscapes should be more than a blur of passing trucks and road signs. Your team is changing—enjoy the process.
- Don't be afraid to backtrack and reflect on the journey. Sometimes we misread the map, or take the longer route and skip an important turn. The time lost is valuable, but the direction gained is even more so.

The Perfect Engine Drives a Solutions Strategy

Keep the long-term perspective. This journey is part of a larger movement, just as kaizen breakthrough is the enabler of the LeanSigma Transformation, and Design for LeanSigma revolutionized process and product design for increased responsiveness and mass customization. Each element of this transformation speaks to a larger mission, and by keeping the bigger objective in mind, we speed through the tougher challenges. In each decade since the Industrial Revolution, manufacturers have been the innovators, the pioneers, indeed the bellwether of methodologies that set the tone for whole economies.

The vision of enterprise-wide Distributive Manufacturing and the discipline of LeanSigma, based on responsive, actual demand-based manufacturing system design, is what will enable companies with roots in the eighteenth and nineteenth centuries to build for the twenty-first and twenty-second centuries.

This is a long-awaited window of opportunity, and it is our responsibility to be prepared, to have the best workers, the best tools, and the energetic leadership required because an opportunity like this comes only once in every one hundred and fifty years.

Exploit it. Good luck.

List of Figures

Figure Number	Figure Title
Figure 1-1	The typical operator's work cycle typically involves a lot of wasted motion and extra walking, as shown in this "spaghetti diagram."
Figure 1-2	The Corporate Challenge is to satisfy the disparate needs of every group served: employees, customers, value chain partners, and shareholders. If the needs of customers, employees, and value chain partners are met, investors will also be pleased.
Figure 1-3	In the LeanSigma Transformation model, we illustrate how the modules of Design for LeanSigma, the Lean Production System, and LeanSigma Value Chain are linked by process improvements and supported on a foundation of Six Sigma process capability, kaizen breakthrough methodology, and senior management leadership.
Figure 1-4	The Maytag Repairman—also known as Ol' Lonely—is symbolic of Maytag's high quality. Innovation is the heart of the company.
Figure 1-5	A value chain map, showing the collaborative efforts of Maytag and Fleetwood RV toward an e-enabled replenishment system. The plants may be

List of Figures

	on opposite sides of the country, but information and product flow is immediate.
Figure 1-6	A sampling of Pella's improvement charts shows lead time dropping by 56 to 65 percent, work-in-process inventory turns going from 12 to 50, and an average annual productivity improvement of 11.3 percent since 1992.
Figure 3-1	Emerging leadership model.
Figure 4-1	Organization charts—before, during, and after transformation.
Figure 4-2	Change methodology.
Figure 4-3	Impact of management enthusiasm.
Figure 4-4	Impact of management.
Figure 4-5	Simple guidelines followed during an assessment.
Figure 4-6	A sample value chain map. This one defines aluminum processing.
Figure 4-7	This Colorgraph shows the Group Dynamics Profile for TBM.
Figure 4-8	LeanSigma Transformation model.
Figure 4-9	LeanSigma progression at work through this process.
Figure 4-10	Key performance measurements.
Figure 5-1	The Lean Production System is supported by the twin pillars of just-in-time and jidoka, resting on a foundation of production smoothing.
Figure 5-2	One-piece flow production versus batch and "fake flow."
Figure 5-3	Example of takt time calculation chart.
Figure 5-4	In this illustration of jidoka, the press is equipped with a light sensor to detect the presence of sheet steel and a second sensor, above the output stack, senses when the tray is full.
Figure 5-5	To control fluctuations in production, we assess monthly demand by product and create a daily schedule that provides every product in sync with the customer's daily volume demands.

List of Figures

Figure 5-6	Lean Production System model line.
Figure 5-7	Japanese characters meaning Continuous Improvement.
Figure 5-8	Description of kaizen methodology.
Figure 5-9	In this example, four operators are working on a process with a sixty-second takt time. Each operator's actual cycle is much shorter than takt, however, with operator D needing only thirty-nine seconds to finish his job. We can focus a kaizen on this issue and possibly reduce the number of operators.
Figure 5-10	Chart showing work sequence.
Figure 5-11	This chart reveals the correct layout of the work cell, a complete parts inventory showing everything required to build the product, and each step the operator takes within a cycle. A series of quality checks is also illustrated in line drawings.
Figure 5-12	Process capacity table.
Figure 5-13	This diagram shows the locations for standard WIP in a process: between each operator's hand-off points, at each automatic machine, and where any time-based processing element, such as curing or drying, takes place.
Figure 5-14	Pease teams quickly achieved outstanding results.
Figure 5-15	Comparative analysis.
Figure 5-16	LeanSigma project phases.
Figure 5-17	For every issue, the team must sketch out multiple possibilities.
Figure 6-1	*The Proportions of the Human Figure* by Leonardo da Vinci, 1490.
Figure 6-2	Machine load and unload.
Figure 6-3	The worker is in a comfortable neutral body position with a slight bend in the hips and the box held close to his body.
Figure 6-4	An illustration of smart materials presentation.
Figure 6-5	(1) Incorrect hand positioning. (2) A small change in hand positioning saves the worker from injury.

List of Figures

	(3) Redesigned tool keeps wrist and hand in right position.
Figure 6-6	Combination tool.
Figure 6-7	The original carrier (photo 1) was difficult for the operator to use. But in photos 2 and 3, the operators rebuilt the carriers with lighter material and hydraulics, making the carriers easier to use.
Figure 6-8	The original workplace design (photos 1, 2,and 3) was stressful on the operator's body. In photo 4, the new design keeps the operator from doing heavy lifting.
Figure 6-9	Parts were kept in large supply racks—an arrangement that required the waterspider to reach into boxes and to lift and bend below waist level (photo 1). In photo 2, the waterspider's replenishment route is changed, freeing up space and saving time.
Figure 7-1	Traditional approach to process design.
Figure 7-2	Elements of design for LeanSigma.
Figure 7-3	Kaizen is used from concept through production.
Figure 7-4	Each module of Design for LeanSigma is rooted in customer desires.
Figure 7-5	Using managed creativity, we consider every alternative in order to find the best concept.
Figure 7-6	Critical evaluation process.
Figure 7-7	Key concepts in Design for LeanSigma.
Figure 7-8	For *each* process step . . . develop a minimum of seven alternatives.
Figure 7-9	A cardboard mock-up of the cell is produced.
Figure 7–10	From the cardboard mock-up, the actual machine is put in place at Pella.
Figure 8-1	Sustaining the gain is a three-pronged approach.
Figure 8-2	The Comprehensive Project List will help prepare for the Senior Management Leadership meeting as well as create common understanding among participants.

List of Figures

Figure 8-3	This directional alignment chart ensures that the company stays on the correct future path.
Figure 8-4	First-year targets should be 20 percent higher than the actual goal.
Figure 8-5	Performance measurements.
Figure 8-6	In the "dumbbell" model, extra raw material and work-in-process are required to cover unreliability in the supply chain and chaotic manufacturing schedules. Extra finished goods build up to cover unreliable replenishment and fluctuations in the market. Once a lean company develops a reliable supply chain and connects with real customer demand, expensive inventory is reduced.
Figure 8-7	Customer service focus.
Figure 9-1	Our wake-up call.
Figure 9-2	In the LeanSigma Value Chain, the twin pillars of supplier development and lean distribution are rooted in business planning and control.
Figure 9-3	Value chain example—retail, current state.
Figure 9-4	Value chain vision—increase value to the customer.
Figure 9-5	Gap analysis, including team assignments.
Figure 9-6	Monthly planning activities.
Figure 9-7	Benefits accumulate throughout the transformed Web-enabled enterprise.
Figure 9-8	Metrics used by Maytag to maintain and reinforce the e-process.

Bibliography

Baghdadi, Me road, Stephen Coley, and David White, *The Alchemy of Growth,* Reading, Mass.: Perseus Books, 1999.

Clausing, Don, *Total Quality Development World Class Concurrent Engineering,* New York: Amacom, 1997.

Fingar, Peter, Harsha Kumar, and Tarvun Sharma, *Enterprise E-Commerce,* Tampa, Fla.: Megham-Kiffer Press, 2000.

Ford, Henry, *Today and Tomorrow,* Portland, Ore.: Productivity Press, first published 1926.

Hirshberg, Jerry, *The Creative Priority,* New York: HarperBusiness, 1998.

Jonash, Ronald S., and Tom Sommerlatte, *The Innovation Premium,* Reading, Mass.: Perseus Books, 1999.

Laraia, Anthony C., Patricia E. Moody, and Robert W. Hall, *The Kaizen Blitz,* New York: John Wiley and Sons, 1999.

Miller, William B., and Vicki L. Schenk, *All I Need to Know about Manufacturing I Learned in Joe's Garage,* Boise, Idaho: Bayrock Press, 2000.

Monden, Yasuhiro, *The Toyota Production System,* Norcross, Ga.: Engineering and Management Press, 1993.

Moody, Patricia E., and Richard Morley, *The Technology Machine,* New York: The Free Press, 1999.

Moore, Geoffrey A., *Crossing the Chasm,* New York: HarperBusiness, 1995.

Nelson, Dave, Rick Mayo, and Patricia E. Moody, *Powered by Honda,* New York: John Wiley and Sons, 1998.

Nelson, Dave, Patricia E. Moody, and Jonathan Stegner, *The Purchasing Machine,* New York: The Free Press, 2001.

Ohno, Taiichi, *The Toyota Production System: Beyond Large-Scale Manufacturing,* Portland, Ore.: Productivity Press, 1988.

Bibliography

Pine, B. Joseph, et al., *The Experience Economy*, Cambridge, Mass.: Harvard Business School Press, 1999.

Target magazine, "Polaroid Scotland's Lean Revolution," Target Volume 16, No. 2, Assn. for Mfg. Excellence, Wheeling, Ill.

Womack, James, and Daniel T. Jones, *Lean Thinking*, New York: Simon & Schuster, 1996.

Index

Airborne, 235
Alexander Doll Co., 60, 61, 230
Allied Signal, 232
Amazon.com, 52
American approach, 230–231
 combined with Japanese approach, 232
Analyze, 101, 138
Assessment process, change and, 92–98
Average quoted lead time, 220–222

Bad processes, evidence of, 37–38
Barnes&Noble.com, 52
Batesville Casket Co., 260
Beal, David, 140, 142
Benchmarks, supplier, 70–71
Bird, Ronald E., 149
Black & Decker, 32–33, 117, 122
 customer satisfaction, 212
 mockups, use of, 190–191, 192
Black belts, 132–133
Bluelight.com 52
Body positions, neutral, 159–161
Boef, Gene de, 28
Brainstorming, silent, 204, 232

Briatico, Tom, 22, 196, 197
Brown, Herb, 61
Buchheit, Mike, 70
Building Organizational Fitness, Management Methodologies for Transformation and Strategic Advantage (Fukuda), 232
Build-to-order projects, 23–25, 259–260
Bureau of Labor Statistics (BLS), 149
Burnham, Dan, 232
Byrne, Art, 54, 61, 67–68, 218

Capital investment as percentage of sales, 228–229
Celestica, 71
Cell-based lean production, 9–10, 47–48
 ergonomics of, 119
Champion training, 132
Change
 assessment process, 92–98
 creativity and, 98–99
 importance of preparing for, 79–80
 long-term commitment to, 80–81
 maintaining, 99–100

Index

Change (*Cont.*)
  Personalysis, use of, 96–98
  as a positive experience, 80
Change, management's role
  assessment process, 92–98
  importance of, 88–90
  kaizen promotion office (KPO), 91–92
  resistance to, 90–91
  visions, developing clear, 87
Change, managing
  cost accounting system, 82–83
  incentive systems, 83
  information systems, 81–82
  organizational structure, 84–86
  supply chain relationships, 82
  workforce involvement, 86–87
Christensen, Gary, 31–32, 54, 74–75, 220, 221
Chrysler Corp., 43–46
Citizen Watch Co., 160
Clausing, Don, 178
Cold Spring Granite, 138–145
Compensation systems, change and, 83
Competition, staying ahead of, 200–201
Concept development, 182–184
Concurrent engineering, 177–178
Consumer behavior, changing, 7–8
Control, 101, 138
Cost(s)
  See also Financial performance
  accounting system, change and, 82–83
  productivity and, 223–225
Crain, David, 168–169
Crames, Allan, 35–37
Creative Priority, The (Hirshberg), 98
Creativity
  before capital, 229
  change and, 98–99
  examples of, 187–189
  reinforcement techniques that build, 192
Critikon, 60, 161
Culture, changing organization, 67
Customers
  design and, 181–182
  return rate, 214
  satisfaction index and quality, 210–215
Customer service focus, maintaining, 235–236
Custom Kitchen Cabinets, Inc., 3
Cycle time chart, 127

Defect
  first-pass yield rate, 214–215
  per unit, 215
  prevention, 120
Delivery compliance percent, 216
Dell Computer Corp., 261
Deming Circle, 100
Deselection, 65
Design for LeanSigma, 17–18, 106–107
  applications, 178–181
  Black & Decker example, 190–191, 192
  compared with concurrent engineering, 177–178
  compared with traditional design, 177
  concept development, 182–184
  creativity, examples of, 187–189
  creativity, reinforcement techniques that build, 192
  customer needs and, 181–182
  defined, 176
  description of learning experience, 173–176
  for manufacture and assembly, 184
  Maytag example, 196–197

Index

mockups, use of, 190–191
Pella example, 193–196
Polaroid example, 193
problems avoided by using, 181–182
process development, 185–187
reviewing, 182
tips, 189
Vermeer example, 198–199
DHL, 235
Directional alignment, 207–208
Disengagement, 99
Distributive Manufacturing
  electronics industry and, 261
  rethinking design for, 260
  role of, 256–257
  tips, 263–264
  Toyota example, 257–259
  workforce and, 262
Dumbbell effect, 218

Eaton, Robert, 44
E-commerce
  effectiveness of, 51
  fallacies of, 57–59
  lean manufacturing, applying, 52–53
  supply chains, 51–52
Economic benefits of ergonomics, 153–156
EFTC, 261
Electronics industry, 261
Employees, fitting them to the job, 158
Enslow, B., 57–59
Ergonomics, 50–51, 119
  defined, 148
  designing products for customer, 152–153
  designing work for the human body, 151–152
  economic benefits of, 153–156
  guidelines for, 157–164
  at Lantech, 150, 153, 168–170

at Maytag, 150–151, 164–167
multitasking and, 171–172
OSHA regulations, 149–150
principles of, 156–157
safety and, 170–171, 225–226
Eureka! problem, 181, 183

Federal Express, 235–236
Financial performance
  capital investment as percentage of sales, 228–229
  net sales growth, 226–227
  operating income as percentage of sales, 227
  R&D cost as percentage of sales, 227
  working capital as percentage of sales, 229–230
Finished goods inventory, 217–219
Flexibility and responsiveness, 215
  average quoted lead time, 220–222
  delivery compliance percent, 216
  finished goods inventory, 217–219
  raw material inventory, 216–217
  product development lead time, 222–223
  total inventory, 216
  work-in-process material inventory, 217
Flextronics International, 71
Flint, Greg, 144
Flow production, 116–118
Ford, Henry, 33, 50, 60
Ford Motors, River Rouge plant, 109–114
Fukuda, Ryuji, 232, 233–234

Gains, maintaining
  American and Japanese approaches, mixing, 232
  American approach, 230–231
  elements for, 203

275

Index

Gains, maintaining (*Cont.*)
   five years of commitment, 202
   Japanese approach, 231–232
   kaizen promotion office, 203, 209–210
   key elements of operations improvement, 201–202
   management, role of, 200–201, 234
   performance metrics, 203, 210–230
   senior management leadership, 203, 204–209
Gamble, Harding, 19
Gap analysis, 243–245
Gartner Group, 57
Gateway Computer, 211, 261
General Electric (GE), 15
General Motors (GM) Corp., 7
   takt time, managing, 41–43
Giddings, Brian, 92, 194
Green belts, 133
Greulich, Gregg, 21–22

Hallmark.com, 52
Haught, Mel, 28, 62–63, 92, 220
Herr, Mike, 32, 92–93, 94–96
Hewlett-Packard, 8
Hicks, Ron, 169
Hillenbrand, Gus, 221
Hillenbrand Industries, 221–222, 228
Hill-Rom, 33, 188
Hirshberg, Jerry, 98–99
Honda BP, 40
Honda of America, Anna, Ohio plant, 147
Hoxan, 230, 231

IBM, 7, 55, 80, 261
Improve, 101, 138
Incentive systems, change and, 83
Information systems, change and, 81–82
Injuries per hundred associates, 225–226

Internal processes and people, leadership for, 66–69
Internet, 261–262
Inventory
   finished goods, 217–219
   raw material, 216–217
   total, 216
   work-in-process material, 217
i2, 55

Japanese approach, 231–232
   combined with American, 232
JCPenney, 218
Jeffress, Charles, 149–150
Jensen, David, 98
Jidoka, 16
   defined, 119–120
   preventing defects with, 120
   role of, 115
Just-in-time, 16, 49
   flow production, 116–118
   line-of-sight management, 118–119
   principles, 116
   pull system, 118
   role of, 115

Kaizen, 6, 15
   breakthrough experience, 108–114
   role of, 121–124
Kaizen promotion office (KPO), 91–92
   at Maytag, 209–210
   role of, 203
Kanban, 49
Kassling, Bill, 67, 229

Lancaster, Pat, 54, 61–62, 74, 168, 222
Lantech, 5, 6, 13, 60
   customer satisfaction index and quality at, 210–212
   ergonomics at, 150, 153, 168–170
   leadership at, 61–62, 74

Index

policy deployment at, 208–209
product development lead time at, 222–223
Leadership/leaders
 by example, 67–68
 early adopters, 66–67
 hands-on, 106
 how to build, 75–78
 for internal processes and people, 66–69
 Lantech example, 61–62, 74
 learning from, 60
 lessons for future, 76–77
 maintaining gains and role of, 203, 204–209
 for market responsiveness, 72–75
 Maytag example, 63–65, 72–74
 model, 78
 role of, 46–49, 61
 Pella example, 62–63, 74–75
 response to bad, 68–69
 for the supply chain, 69–71
Lead time
 average quoted, 220–222
 product development, 222–223
Lean manufacturing
 applying to e-commerce, 52–53
 cell-based, 9–10
 challenge of, 6–7
 cost of, 49–50
 examples of, 5–6, 8–9
 reasons for, 4
Lean Production System, 16–17
LeanSigma, 6
 Six Sigma versus, 101–103
LeanSigma, designing. *See* Design for LeanSigma
LeanSigma Transformation, 10–15
 phases of, 100–101
 reasons companies don't use, 53–55
 role of, 100

steps/guidelines, 103–106
LeanSigma Value Chain, 18
 map, 95–96
Learmonth, Art, 24–25, 92
Lienenbrugger, Herb, 28
Line-of-sight management, 118–119
Lopez Method, 69
Lutz, Bob, 43–44

Mai, Phung, 137
MAIC wheel, 138
Mallon, William, 154–156
Management
 See also Change, management's role; Leadership/leaders
 line-of-sight, 118–119
 maintaining gains and role of, 200, 203, 204–209, 234
Manufacturing, traditional
 defined, 177
 example of, 1–3
 problems with, 4–5
Marketing, design and, 181–182
Market responsiveness, leadership for, 72–75
Markets, creating new, 213
Marks, Michael, 71
Maruo, Teruyuki, 67
Maytag, 5, 6
 build-to-order projects, 23–25
 Cleveland Cooking Products, 19–25
 delivery compliance percent at, 216
 design at, 196–197
 ergonomics at, 150–151, 164–167
 kaizen promotion office at, 209–210
 leadership at, 63–65, 72–74
 value chain at, 250–255
Measure, 100, 138
Medical costs per hundred associates, 226

277

Index

Mercedes-Benz, 5, 6, 60, 61
   Brazil plant, 25–27, 188–189
Mizuguchi, Keiji, 54
Mockups, use of, 190–191
Model behavior, 68
Moonshining, 185–187
Morin, Tom, 190–191
Motorola Inc., 10, 15, 232
Multitasking, ergonomics and, 171–172
Musculoskeletal disorders (MSDs), 149

Net sales growth, 226–227
Nissan Design International, Inc., 88, 98–99

Objectives, linking transformation to, 103–105
Observation
   evidence of bad processes, 37–38
   learning to see, 35–41
   management, role of, 46–49
Ohno, Taiichi, 33, 60, 114, 119, 130, 233
Operating income as percentage of sales, 227
Operations improvement, key elements of, 201–202
Organizational structure, change and, 84–86
OSHA (Occupational Safety and Health Administration), 149–150
Outpost.com, 52
Outside facilitator, use of expert, 206–207

Parts proliferation, design and, 182
Pease Industries, 133–138
Pella Corp., 5, 6, 13, 60
   average quoted lead time at, 220
   capital reduction at, 228
   delivery compliance percent at, 216
   design at, 193–196
   ergonomics at, 147–148
   kaizen, example of, 27–32
   leadership at, 62–63, 74–75
   maintaining competitiveness, 221, 222
Performance metrics
   building momentum, 213–214
   cost and productivity, 223–225
   customer satisfaction index and quality, 210–215
   financial performance, 226–230
   flexibility and responsiveness, 215–223
   role of, 203
   safety and ergonomics, 225–226
Personalysis, 96–98
Polaroid, 33, 88, 96
   design at, 193
Policy deployment, 208–209
Pratt & Whitney, 121
Process Capacity Table, 128
Procter & Gamble, 7
Product development lead time, 222–223
Product focus teams, 84–86
Production Smoothing, 115–116, 120–121
Production system
   advanced quality tools, 131–132
   champions, black and green belts, 132–133
   components of, 115–116
   getting started example, 133–145
   jidoka, 115, 119–120
   just-in-time, 115, 116–119
   Kaizen Breakthrough experience, 108–114, 121–124
   managing for daily improvement, 130–131

Index

model line, 121
Production Smoothing, 115–116, 120–121
standard operations, 124–130
Productivity, cost and, 223–225
Pull method, 81–82, 118
Push method, 81–82

Quality
  customer satisfaction index and, 210–215
  problems, 100

R&D cost as percentage of sales, 227
Raw material inventory, 216–217
Raytheon, 232
Ritz Carlton, 212

Safety, 158, 170–171, 225–226
Samsonite, 47, 49
Sears, 218
Shainen, Dorian, 33, 210
Shewhart, Robert, 33
Shingijutsu Co., 114
Shingo, Shigeo, 33, 54
Silent brainstorming, 204, 232
Six Sigma, 10, 15–16, 232
  LeanSigma versus, 101–103
Smith, Bonnie, 100, 102–103
Solectron, 71
Sony, 233
SRI, 71
Standard operations
  defined, 124–125
  importance of, 130–131
  necessity of, 125–130
  takt time/cycle time chart, 127
  work in process, 128–130
  work sequence, 127–128
Standard work, 234–235
Standard Work Combination sheet, 128

Standard work in process (SWIP), 128–130
Standard Work Layout sheet, 127–128
Sullivan, Dan, 21, 93, 198
Sundstrand Aerospace, 153
Supply chains
  benchmarks, 70–71
  change and, 82
  communicating with, 69–70
  leadership for, 69–71
  lean, 218–219
  responsiveness, 71
Swoyer, Sam, 24

Takt time
  managing, 41–43, 116–118
  standard operations, 127
Tasks, importance of varying, 163
TBM Consulting Group, 14, 16, 100, 114
Teams, preparation and selection of, 96–98
Time, 51, 52
Tofas, 33
Tool handles, redesigning, 161–162
Total inventory, 216
Toyota Production System, 14, 16, 40, 114
  Distributive Manufacturing at, 257–259
Toysrus.com, 52

Uhrich, Carole, 61, 63–65
Unions, 86–87
Unipart, 32, 220
UPS, 235
U.S. Postal Service, 235–236

Value chain
  business planning and control, 249–250

279

Index

Value chain *(Cont.)*
 developing, 240–242
 fresh ideas and critical criteria, 242–243
 identifying gaps, 243–245
 information systems, role of, 249
 map, 95–96
 Maytag example, 250–255
 model, 248–249
 replenishing system, creating a, 246
 selling the system, 246–247
 supply-side transformation, 247–248
 transforming distribution, 245–246
Value per associate, 224–225
Variability control, 100
Vermeer Manufacturing, 5, 6
 design at, 198–199
Voice-of-the-customer data, 181–182

Wabtec, 229
Wal-Mart, 218
Ward, Lloyd, 72–74
Weingarten, Karsten, 26, 27, 61
Welch, Jack, 15
Winbook, 261
Wiremold, 5, 6, 13, 61
Workforce
 change and the involvement of, 86–87
 Distributive Manufacturing and, 262
Working capital as percentage of sales, 229–230
Work in process (WIP), 128–130
 material inventory, 217
Workplace, fitting employees to, 158–159
Work sequence, 127–128

Xerox, 79
X-type matrix, 232, 233–234

Zarrabi, Ramin, 21, 24, 92

About the Authors

Anand Sharma is the CEO of TBM Consulting Group, an international training and consulting organization that was instrumental in bringing lean business principles to the United States, Europe, and South America over the past decade. He was named a "Hero of Manufacturing" for 2001 by *Fortune* magazine.

Mr. Sharma is a regular columnist for *Global Manufacturing Review*, where he writes about trends in international business. He has been extensively quoted on business, manufacturing, and the new economy in *Time* and *Fortune* magazines, the Dow Jones Newswire, and CNN/FN, as well as other national media.

A former executive with American Standard, Mr. Sharma is an advisor to dozens of corporations around the globe and currently sits on the board of Madame Alexander Doll and TBM Holdings Co. Clients of TBM and Mr. Sharma include business leaders such as Butler Manufacturing, Maytag, Mercedes-Benz do Brazil, Michelin, Pella Corporation, Polaroid, Siemens.

Patricia E. Moody, CMC (PEMoody@aol.com), named by *Fortune* magazine one of "Ten Pioneering Women in Manufacturing," is a manufacturing

About the Authors

management consultant and writer with over thirty years of industry, consulting, and teaching experience. Her eight books include *The Purchasing Machine: How the Top Ten Companies Use Best Practices to Manage Their Supply Chains,* with Dave Nelson (www.purchasingmachine.com); *The Technology Machine: How Manufacturing Will Work in the Year 2020,* with Richard E. Morley; *The Kaizen Blitz;* and *Powered by Honda.*

Her client list includes industry leaders Cisco, Honda, Solectron, Motorola, Johnson & Johnson, Mead Corporation, and Harvard Business School Publishing.

As former editor of AME's *Target* and a frequent contributor to *iSource, Sloan Management Review, Manufacturing Asia,* and other manufacturing and supply chain media, she creates breakthrough work on the future of manufacturing, teams, kaizen, new product development, supply management, and e-commerce and e-manufacturing[sm]. She is a member of the Sloan Management Review Editorial Advisory Board and the Board of the Institute for Supply Chain Management, is a contributing editor for *Supply Strategy,* and has appeared in various international media, including CNN's "21st Century" with Bernard Shaw, KQED's "Tech Nation," and interviews in *Fortune, Manufacturing Asia,* and *Purchasing.*

Printed in the United States
By Bookmasters